Nelson Advanced Science

Periodicity, Quantitative Equilibria and Functional Group Chemistry

revised edition

Rod Beavon • Alan Jarvis

endorsed by
edexcel

Nelson Thornes
a Wolters Kluwer business

First published in 2001 by:
Nelson Thornes Ltd
Delta Place
27 Bath Road
CHELTENHAM
GL53 7TH
United Kingdom

This edition published in 2003

06 07 / 10 9 8 7 6 5 4 3

A catalogue record for this book is available from the British Library

ISBN 0 7487 7657 5

Illustrations by Hardlines and Wearset
Page make-up by Hardlines and Wearset

Printed in Croatia by Zrinski

Acknowledgements
The authors and publisher are grateful for permission to include the following copyright material:

Examination questions are reproduced by permission of Edexcel.

Photographs:
Corbis: 2.1 Bettmann;
Getty Stone: 4.1 Robert Frerck;
Getty Telegraph: cover Greg Pease;
Nelson Thornes: 4.2, 4.3 Andy Ross, 4.7 Stuart Boreham, 4.4 Alan Thomas;
Royal Society of Chemistry: 5.3;
Science Photolibrary: 2.3 Alex Bartell, 1.2, 1.3, 5.2;
Science & Society Picture Library: page vii.

Every effort has been made to trace all the copyright holders, but where this has not been
possible the publisher will be pleased to make any necessary arrangements at the first
opportunity.

Contents

Introduction

This series has been written by the Chief Examiner and others involved directly with the development of the Edexcel Advanced Subsidiary (AS) and Advanced (A) GCE Chemistry specifications.

Periodicity, Quantitative Equilibria and Functional Group Chemistry is one of four books in the Nelson Advanced Science (NAS) series developed by updating and reorganising the material from the Nelson Advanced Modular Science (AMS) books to align with the requirements of the Edexcel specifications from September 2000. The books will also be useful for other AS and Advanced courses.

Periodicity, Quantitative Equilibria and Functional Group Chemistry provides coverage of Unit 4 of the Edexcel specification. The book builds on ideas that have been introduced in the AS units. Hess's Law is extended to the structure of crystals, and the implications for the properties of ionic substances discussed. Further Inorganic Chemistry appears with more detailed considerations of Period 3 and Group 4, reinforcing and amplifying trends in the Periodic Table. Chemical Equilibrium, one of the most important unifying themes in chemistry, is developed quantitatively and extended to the theory of acids and bases. Further Organic Chemistry is covered to enable synthetic routes to be developed, and to enhance problem-solving skills in chemical synthesis.

Other resources in this series

NAS *Teachers' Guide for AS and A Chemistry* provides advice on safety and risk assessment, suggestions for practical work, key skills opportunities and answers to all the practice and assessment questions provided in *Structure, Bonding and Main Group Chemistry*; *Organic Chemistry, Energetics, Kinetics and Equilibrium*; *Periodicity, Quantitative Equilibria and Functional Group Chemistry*; and *Transition Metals, Quantitative Kinetics and Applied Organic Chemistry*.

NAS *Make the Grade in AS and A Chemistry* is a Revision Guide for students. It has been written to be used in conjunction with the other books in this series. It helps students to develop strategies for learning and revision, to check their knowledge and understanding and to practise the skills required for tackling assessment questions.

Features used in this book

The Nelson Advanced Science series contains particular features to help you understand and learn the information provided in the books, and to help you to apply the information to your coursework.

These are the features that you will find in the Nelson Advanced Science Chemistry series:

Text encapsulates the necessary study for the Unit. Important terms are indicated in **bold**.

5 Oxidation/reduction: an introduction

Introduction

Oxidation and reduction are found with all but four elements in the Periodic Table, not just with the transition metals, although they show these reactions to such an extent that they could be accused of self-indulgence.

When magnesium reacts with oxygen (Figure 5.1)

$$2Mg(s) + O_2(g) \rightarrow 2MgO(s)$$

the product contains Mg^{2+} and O^{2-} ions. Reaction with oxygen is pretty clearly oxidation. The reaction of magnesium with chlorine

$$Mg(s) + Cl_2(g) \rightarrow MgCl_2(s)$$

MNEMONIC

OIL RIG:

oxidation **i**s **l**oss

reduction **i**s **g**ain

gives a compound with Mg^{2+} and Cl^- ions. In both cases the magnesium atom has lost electrons, so as far as the magnesium is concerned the reactions are the same. This idea is generalised into the definition of oxidation as loss of electrons. Reduction is therefore the gain of electrons. Since electrons don't vanish from the universe, oxidation and reduction occur together in **redox** reactions.

Oxidation numbers

For simple monatomic ions such as Fe^{2+} it's easy to see when they are oxidised (to Fe^{3+}) or reduced (to Fe). For ions such as NO_3^- or SO_3^{2-} which also undergo oxidation and reduction it is not always so easy to see what is happening in terms of electrons. To assist this, the idea of **oxidation number** or **oxidation state** is used. The two terms are usually used interchangeably, so that an atom may have a particular oxidation number or be in a particular oxidation state.

Fig. 5.1 The use of magnesium flares in photography being demonstrated at an early meeting of the British Association in Birmingham (1865).

Definition boxes in the margin highlight some important terms.

INTRODUCTION

Questions in the margin will give you the opportunity to apply the information presented in the adjacent text.

The **empirical formula** shows the ratio of atoms present in their lowest terms, i.e. smallest numbers. Any compound having one hydrogen atom for every carbon atom will have the empirical formula CH; calculation of the **molecular formula** will need extra information, since ethyne, C_2H_2, cyclobutadiene, C_4H_4, and benzene, C_6H_6, all have CH as their empirical formula. Empirical formulae are initially found by analysing a substance for each element as a percentage by mass.

> **QUESTION**
>
> Find the empirical formula of the compound containing C 22.02%, H 4.59%, Br 73.39% by mass.

Practice questions are provided at the end of each chapter. These will give you the opportunity to check your knowledge and understanding of topics from within the chapter.

Assessment questions are found at the end of the book. These are similar in style to the assessment questions for Advanced GCE that you will encounter in your Unit Tests (exams) and they will help you to develop the skills required for these types of questions.

Questions

1 Plot on a (small) graph the first ionisation energies of the elements from sodium to argon, and account for the shape obtained.

2 Use data from a data book to plot a graph of atomic radius vs atomic number for the elements of Periods 2 and 3 (Li to Ar). Account for the difference in the graphs between Groups 2 and 3.

3 Sketch the structures of:
  (a) the giant covalent lattice of silicon
  (b) the molecule P_4
  (c) the molecule S_8.

4 Silicon has no compounds in which the silicon atom forms double bonds with other elements. Phosphorus, by contrast, does form double bonds with other elements. Suggest why silicon and phosphorus are different in this respect.

Acknowledgements

I want to thank the following most sincerely for invaluable help and advice: at Westminster, Peter Hughes, Derek Stebbens, Martin Robinson, Gilly French, Damian Riddle and Nick Hinze; at Edexcel, Ray Vincent; and of course Alan Jarvis, for his friendship and advice during the years we worked together. My students have taught me more than almost anyone else – to them I owe a particular debt. Thanks too to my late wife Doreen, who was so supportive during the initial writing.

The authors and publisher would like to thank Geoff Barraclough for his work as Series Editor for the original series of four NAMS books, from which the new suite of NAS books was developed.

About the Authors

Rod Beavon is Chief Examiner in Chemistry for Edexcel and Head of Science at Westminster School, London.

The late **Alan Jarvis** was former Head of Chemistry at Stoke-on-Trent Sixth Form College and was Chief Examiner in Chemistry for Edexcel.

Ionic crystals

Many people admire crystals – their symmetry, and often bright colours, have appealed throughout the ages. The symmetry reflects the order found at the atomic level, and a study of the energy changes during their formation can give information about their bonding.

Why does sodium react with chlorine to give sodium chloride? The simple 'explanation' says that in doing so both sodium and chlorine achieve, in their ions, a noble gas electronic structure. This is of course true; but it is only part of the story. The production of the ions is, overall, endothermic — and it is the bringing together of these separate ions into the solid crystal that is exothermic enough to compensate.

Lattice energy

The first ionisation energy for sodium, which is the energy change per mole for:

$$Na(g) \rightarrow Na^+(g) + e^-$$

and the electron affinity of chlorine, which is the energy change per mole for the process:

$$Cl(g) + e^- \rightarrow Cl^-(g)$$

are significant here, together with the enthalpy of formation of sodium chloride, defined as the enthalpy change per mole for:

$$Na(s) + \tfrac{1}{2}Cl_2(g) \rightarrow NaCl(s)$$

These can be combined using Hess's Law, together with the enthalpies of atomisation of the elements; here, the significance of the lattice energy, ΔH_{latt}, becomes evident. First we shall remind ourselves about Hess's Law and define lattice energy.

Hess's Law revisited

In Unit 2 Hess's Law was introduced. The heat energy change (enthalpy change) for a reaction depends only on the *initial* and *final states* of the system, and not on the pathway between those states. In Unit 2, we used Hess's law to calculate enthalpy changes for reactions which cannot be found by experiment, by combining the enthalpy changes of other, possible, reactions. The use of enthalpies of formation and of combustion in this way will be familiar.

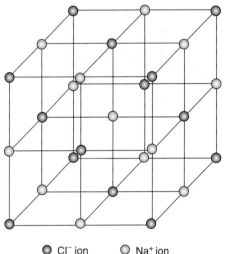

○ Cl⁻ ion   ○ Na⁺ ion

Fig. 1.1 The structure of sodium chloride.

IONIC CRYSTALS

> **DEFINITION**
>
> Lattice energy, ΔH_{latt}:
> energy change per mole for
> the process
> $M^+(g) + X^-(g) \rightarrow MX(s)$

Fig. 1.2   Fritz Haber (1868–1934) invented the Haber process for the production of ammonia from nitrogen, and contributed to crystal theory via quantum mechanics.

Fig.1.3   Max Born (1882–1970) dealt with the theory of crystal lattices (1912) before their reality had been demonstrated.

The lattice energy

The lattice energy of sodium chloride is defined as the energy change per mole for the process

$$Na^+(g) + Cl^-(g) \rightarrow NaCl(s)$$

In general, it is the energy change for the formation of one mole of the solid crystal from the gaseous ions. This definition means that lattice energies are *exothermic*; many books define this quantity in the endothermic direction, so make sure you know which convention is being used.

Lattice energies are large. It is this large energy release that enables the reaction of sodium and chlorine to take place, and this is shown in the Born–Haber cycle (see Figure 1.4).

The lattice energy cannot be experimentally determined; it can be found only by using the Born–Haber cycle. Nevertheless this value is called the experimental value of ΔH_{latt} since the Born–Haber cycle uses experimentally determined values. The value thus obtained is the lattice energy which the compound actually possesses; it is the real lattice energy. The distinction between this and the theoretical or calculated lattice energy is discussed on page 8.

Enthalpy changes

The enthalpy changes involved in the Born–Haber cycle can be defined more formally (and generally):

- $\Delta H_a^{\ominus}$, the standard enthalpy of atomisation, is the enthalpy change for the production of one mole of atoms in the gas phase from the element in its standard state at 1 atmosphere pressure and the stated temperature. Note that it is defined per mole of atoms formed;

- $I(1)$, the first ionisation energy, is the energy change for the conversion of one mole of gaseous atoms into one mole of positive ions in the gas phase;

$$M(g) \rightarrow M^+(g) + e^-$$

- The second ionisation energy is defined similarly, being the energy change per mole for the process

$$M^+(g) \rightarrow M^{2+}(g) + e^-$$

- E.A.(1) is the first electron affinity, which is the energy change for the conversion of one mole of gaseous atoms into one mole of negative ions in the gas phase;

$$X(g) + e^- \rightarrow X^-(g)$$

- The second electron affinity is defined similarly:

$$X^-(g) + e^- \rightarrow X^{2-}(g)$$

- $\Delta H_f^{\ominus}$, the standard enthalpy of formation, is the enthalpy change accompanying the formation of one mole of compound from its elements, all substances being in their standard states at 1 atm pressure and a defined temperature. Unless otherwise stated, the temperature is 298 K – but this value does not form part of the definition.

All of the above can be obtained experimentally, either directly or, in some cases, indirectly via other Hess's Law cycles.

Watch that sign: a source of errors

Many errors in thermochemical calculations can be laid at the door of the simple signs + and –. The problem is that they are used in several ways:

- as an instruction: a + b means 'add the number b to the number a'. This use of +, as an instruction, is called a binary operation since it is an operation on two things. Other binary operations include those represented by –, ×, and ÷.

- as an indication of whether a number is greater than zero (e.g. +12) or less than zero (e.g. –2).

- and, in chemistry, as a convention, to indicate the direction of movement of heat; + into the system, and – out of the system.

Errors can be reduced dramatically if you distinguish, in your calculations, between the instruction to add or subtract, and the sign convention; this is done in the examples below by using brackets to separate the instruction + (add) from the convention + (endothermic).

The Born–Haber cycle

The Born–Haber cycle shows the relationship between enthalpy changes and lattice energy. It can be shown in two ways: as an energy-level diagram (Figure 1.4), or as a cycle (Figure 1.5).

In the energy-level diagram for the Born–Haber cycle for sodium chloride, each horizontal line represents the energy level. The species existing on this level are written on the line together with their states. The gaps represent the enthalpy changes accompanying the change in the species shown, and can be drawn to scale if desired.

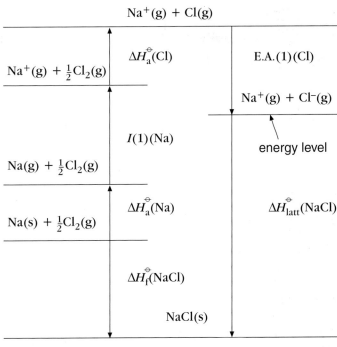

Fig. 1.4 The energy-level form of the Born–Haber cycle for sodium chloride.

The cycle form for sodium chloride is shown in Figure 1.5 below.

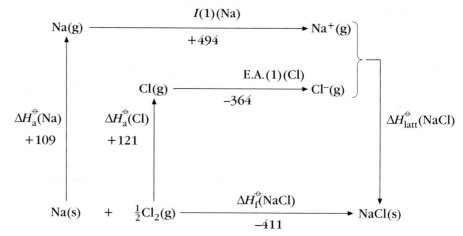

Fig. 1.5 The cycle form of the Born–Haber cycle for sodium chloride.

From both of these diagrams we have:

$$\Delta H_f^\ominus(NaCl) = \Delta H_a^\ominus(Na) + \Delta H_a^\ominus(Cl) + I(1)(Na) + E.A.(1)(Cl)$$
$$+ \Delta H_{latt}^\ominus(NaCl)$$

Substituting values in kJ mol^{-1}, and remembering to use brackets to distinguish signs, we obtain:

$$(-411) = (+109) + (+121) + (+494) + (-364) + \Delta H_{latt}(NaCl)$$

from which

$$\Delta H_{latt}(NaCl) = -771 \text{ kJ mol}^{-1}$$

Factors affecting the magnitude of the lattice energy

The lattice energy depends on:

- the sizes of the ions;

- their charges;

- the crystal structure of the compound;

- the extent to which the bonding deviates from the fully ionic model.

The lattice of an ionic crystal is held together by a balance of attractive forces between ions of opposite charge and repulsive forces between those of like charge; overall there is net attraction.

The force between a pair of ions of charge z^+ and z^- with radii r_+ and r_- is given by:

$$F = \frac{z^+ z^-}{(r_+ + r_-)^2}$$

From this it is clear that the lattice energy will increase:
- as the magnitudes of the charges z^+ and z^- increase
- as the distance $(r_+ + r_-)$ between the ions decreases.

Lattice energies are therefore larger for compounds of small Group 2 cations with small anions, than for larger Group 1 cations with large anions. Experimental values for the lattice energies of the halides of Group 1 and the chlorides and oxides of Group 2 are given in Tables 1.1 and 1.2.

Table 1.1 *Experimental lattice energies in kJ mol^{-1} for the alkali metal halides*

	Fluoride	Chloride	Bromide	Iodide
lithium	−1031	−848	−803	−759
sodium	−918	−780	−742	−705
potassium	−817	−711	−679	−651
rubidium	−783	−685	−656	−628
caesium	−747	−661	−635	−613

The alkali metal halides show the effect of increasing size of cation or anion; ΔH_{latt} decreases in magnitude from LiF to CsF, or from LiF to LiI, since in both cases the distance between the ions is increasing. The strongest lattice is LiF, with the smallest ions; the weakest is CsI, with the largest. A word of caution, however; our comparisons are less quantitative than they might appear. The comparisons would require that all the crystal structures considered are the same, but they are not. **The lattice energy depends on the crystal structure as well as on the ions, since the distance between the ions depends on the structure that is adopted.**

ΔH_{latt} would be expected to be larger for Group 2 halides, since the cation is more highly charged and smaller than Group 1, and even larger for oxides, where both ions have a double charge and the radius of the oxide ion (140 pm) is less than that of the chloride ion (180 pm). Table 1.2 shows that both these predictions are true.

Table 1.2 *Experimental lattice energies in kJ mol^{-1} for Group 2 chlorides and oxides*

$MgCl_2$	−2493	MgO	−3889
$CaCl_2$	−2237	CaO	−3513
$SrCl_2$	−2112	SrO	−3310
$BaCl_2$	−2018	BaO	−3152

ΔH_{latt} and experiment

It was stated above that ΔH_{latt} cannot be determined experimentally. It is reasonable to ask why not; surely if NaCl were heated sufficiently, gas phase ions would be produced? Alas, it is not so; sodium chloride vapour at temperatures not much above its boiling temperature of 1413°C consists of ion pairs Na^+Cl^-. Increasing the temperature causes dissociation into *atoms* Na and Cl, and not into *ions*. Indeed the high lattice energy of sodium chloride would lead to a much higher boiling temperature than it actually has if these ion pairs were not formed. On boiling, not all the interionic forces found in the solid state are broken.

Applying the lattice energy

Lattice energies are useful in considering a variety of problems:

- the stoichiometry of salts (for example why magnesium chloride is $MgCl_2$ and not $MgCl$);

- the solubility of ionic compounds;

- the thermal decomposition of salts.

The first two of these will now be considered (thermal decomposition of salts is covered in Unit 1).

The stoichiometry of salts

The word *stoichiometry*, referring to the ratios in which atoms combine or react, comes from Greek, meaning *element measurer*. Students often ask: why isn't the formula for magnesium chloride $MgCl$ rather than $MgCl_2$?

The answer can be found from the Born–Haber cycle. **All chemical bonding occurs because the system of bound atoms (a molecule, or an ionic or metallic lattice) is of lower energy than the separated atoms**. The Born–Haber cycles for $MgCl$ and $MgCl_2$ show that the latter system has the *lower energy* of the two. We have to find the enthalpy of formation of $MgCl$, that is ΔH_f for

$$Mg(s) \quad + \quad \tfrac{1}{2}Cl_2(g) \quad \rightarrow \quad MgCl(s)$$

The lattice energy can be estimated; Na^+ and Mg^+ would be expected to be of similar size, so the crystal structure of $MgCl$ and $NaCl$ would be the same and ΔH_{latt} similar too. Figures 1.6 and 1.7 show energy level diagrams for Born–Haber cycles relating to $MgCl$ and $MgCl_2$, using the following data:

	$\Delta H^{\ominus}$/ kJ mol^{-1}
$Mg(s) \rightarrow Mg(g)$	+150
$Mg(g) \rightarrow Mg^+(g) + e^-$	+736
$Mg^+(g) \rightarrow Mg^{2+}(g) + e^-$	+1450
$Cl_2(g) \rightarrow 2Cl(g)$	+242
$Cl(g) + e^- \rightarrow Cl^-(g)$	–364
$Mg^+(g) + Cl^-(g) \rightarrow Mg^+Cl^-(s)$	–770
$Mg^{2+}(g) + 2Cl^-(g) \rightarrow Mg^{2+}(Cl^-)_2(s)$	–2493

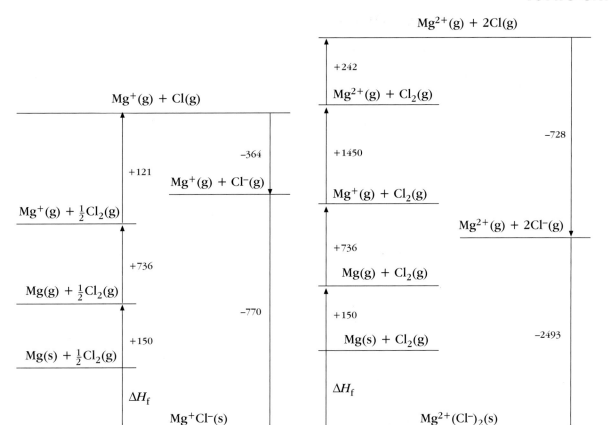

Fig. 1.6 *The Born–Haber cycle for MgCl.*   Fig. 1.7 *The Born–Haber cycle for MgCl₂.*

Applying Hess's law to the cycle in Figure 1.6 gives:

$$\Delta H_f^{\ominus}(\text{MgCl}) = (+150) + (+736) + (+121) + (-364) + (-770)$$

$$\Delta H_f^{\ominus}(\text{MgCl}) = -127 \text{ kJ mol}^{-1}.$$

Similarly for Figure 1.7, the calculation is:

$$\Delta H_f^{\ominus}(\text{MgCl}_2) = (+150) + (+736) + (+1450) + (+242) + 2(-364)$$
$$+ (-2493)$$

$$\Delta H_f^{\ominus}(\text{MgCl}_2) = -643 \text{ kJ mol}^{-1}.$$

Both enthalpies of formation are exothermic, but the formation of $MgCl_2$ is five times as energetically favourable as MgCl. MgCl is never formed. The main factor which favours the formation of $MgCl_2$ is the very high lattice energy. This more than compensates for the additional energy that has to be supplied for the second ionisation of magnesium to form Mg^{2+} from Mg^+.

Theoretical lattice energies

An ionic crystal possesses only one lattice energy, and that is the experimental value already mentioned. A real crystal like that of sodium chloride will not be fully ionic; there will always be some degree, even though it may be very little, of covalent bonding due to electron sharing. So it is instructive to compare the real lattice energy obtained from a Born–Haber cycle with one calculated from a model where the crystal is considered to be completely ionic.

The calculation of the theoretical lattice energy involves extending Coulomb's Law to three dimensions, considering repulsions as well as attractions and taking account of the crystal structure. Comparison of the theoretical lattice energy with the experimental value gives a measure of the deviation of the crystal from the ionic model. The deviations, which make the crystal stronger than the ionic model predicts, give an indication of the contribution of covalent bonding. Deviations would be expected to be largest where the ions are small and of high charge, i.e. the most polarising.

Changing the cation size

Table 1.3 shows the effect of changing the cation size on the difference between experimental (Born–Haber) and theoretical lattice energies for the chlorides of Groups 1 and 2.

Table 1.3 *Experimental (B–H) and theoretical lattice energies (in kJ mol^{-1}) for the chlorides of the s-block*

	Cation radius/pm	Lattice energy (B–H)	Lattice energy (theory)		Cation radius/pm	Lattice energy (B–H)	Lattice energy (theory)
NaCl	102	−780	−770	$MgCl_2$	72	−2526	−2326
KCl	138	−711	−702	$CaCl_2$	100	−2258	−2223
RbCl	149	−685	−677	$SrCl_2$	113	−2156	−2127
CsCl	170	−661	−643	$BaCl_2$	136	−2056	−2033

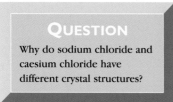

QUESTION

Why do sodium chloride and caesium chloride have different crystal structures?

It would be expected that the greatest deviations from the ionic model would be for the smallest, most polarising cations. It is certainly true for $MgCl_2$ compared with the others. But how do we interpret the values for the Group 1 chlorides? The question really is, are they so very different? Bearing in mind that both the experimental and the theoretical values will have an error associated with them (in one case experimental error and in the other errors inherent in the theoretical model used), the evidence for these chlorides is that they are virtually completely ionic. Furthermore, CsCl does not have the same crystal lattice as the others, and although this will have been taken into account in the calculation, the errors may very well not be the same.

Changing the anion size

If the anion size changes, we expect greater deviation from the ionic model with the larger, more polarisable anions. Iodides should therefore show greater differences than fluorides; Table 1.4 shows that this is true.

Table 1.4 *Theoretical and experimental lattice energies (in kJ mol⁻¹) for sodium and magnesium halides*

	Anion radius/pm	Lattice energy (B–H)	Lattice energy (theory)		Lattice energy (B–H)	Lattice energy (theory)
NaF	136	−918	−912	MgF$_2$	−2957	−2913
NaCl	181	−780	−770	MgCl$_2$	−2526	−2326
NaBr	195	−742	−735	MgBr$_2$	−2440	−2097
NaI	216	−705	−687	MgI$_2$	−2327	−1944

Lattice energy and the solubility of ionic compounds

A simple theory of solubility for ionic compounds, based solely on enthalpy changes, cannot be stated. This is because the solubility of most substances is not controlled by enthalpy changes alone, any more than the direction of spontaneous change is so controlled in chemical reactions. The problem cannot be examined generally without using the thermodynamic idea of entropy, and unfortunately quite small changes in entropy can have large consequences for solubility. This is beyond our present concerns, so we shall look only at compounds for which the enthalpy change of solution, ΔH_{soln}, is the governing factor.

When such an ionic substance dissolves, the enthalpy change depends upon:

- the lattice energy, ΔH_{latt}, of the solid
- the hydration enthalpy, ΔH_{hyd}, of the ions

ΔH_{hyd} is defined as the heat change per mole for the hydration of the gaseous ion with enough water for there to be no further heat change on dilution. For a unipositive cation ΔH_{hyd} is exothermic, for example:

$$M^+(g) + aq \rightarrow M^+(aq)$$

If ΔH_{hyd} is roughly the same magnitude as ΔH_{latt}, or is greater, then the heat needed to break the lattice is recouped by the hydration of the ions, and the dissolution of the solid is favoured.

In this case ΔH_{soln} is negative. A low solubility may be a consequence of high lattice energy or a low hydration enthalpy of the ions.

> **DEFINITION**
>
> ΔH_{hyd} is the heat change per mole for the hydration of the gaseous ion with enough water for there to be no further heat change on dilution.

The enthalpy changes for a salt M^+X^- can be represented on the Hess's Law cycle:

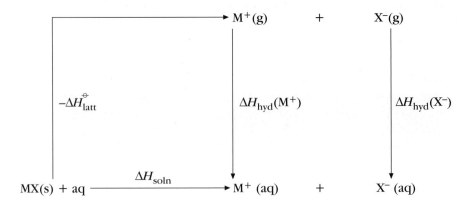

Fig. 1.8 *Hess's Law cycle for the salt M^+X^-.*

From Hess's Law,

$$\Delta H_{soln} = \Delta H_{hyd}(M^+) + \Delta H_{hyd}(X^-) - \Delta H_{latt}(M^+X^-)$$

The hydration of ions occurs because the polar water molecules are attracted to the charge on the ion (Figure 1.9). The extent of the hydration depends on the charge density of the ion, i.e. its charge per unit surface area, so the higher the charge and the smaller the ion, the more exothermic the hydration enthalpy will be.

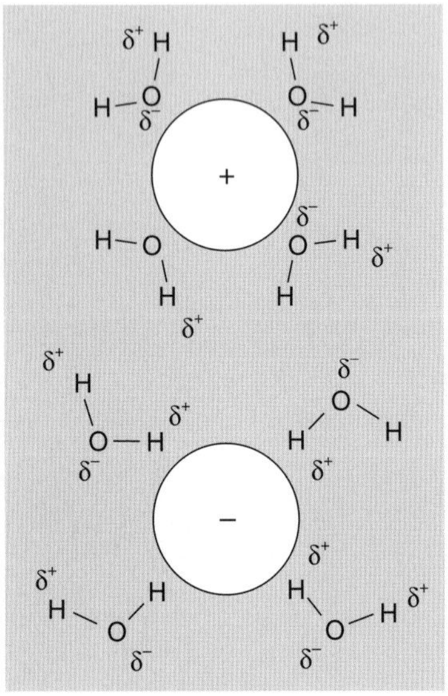

Fig. 1.9 The hydration of ions – polar water molecules are attracted to the charge on the ion.

Solubility trends in the hydroxides and sulphates of Group 2

Trends in solubility of a given type of compound for different metals in a group depend on which of the changes, in hydration enthalpy of the cation or in the lattice energy of the compounds, are more important as the cation size increases down the group.

In the case of the sulphates of Group 2, the hydration enthalpy of the cation has most effect. The lattice energy is partly a function of the sum of the radii of the cation and anion (see page 4). The size of the sulphate ion is such that this quantity does not change very much as the cation size changes. Therefore the contribution from ΔH_{latt} is similar for all of the Group 2 sulphates. However, ΔH_{hyd} of the cations falls significantly as they get larger. Therefore the solubility of the sulphate falls with increasing cation size, because the lattice energy is not exceeded so much by the hydration enthalpy.

Table 1.5 *Solubility of Group 2 sulphates and hydroxides*

	Hydration enthalpy/kJ mol^{-1}	Cation radius /pm	Sulphate/ mol per 100 g water	Hydroxide/ mol per 100 g water
Mg^{2+}	−1920	65	1.83×10^{-1}	2.00×10^{-5}
Ca^{2+}	−1650	99	4.66×10^{-3}	1.53×10^{-3}
Sr^{2+}	−1480	113	7.11×10^{-5}	3.37×10^{-3}
Ba^{2+}	−1360	135	9.43×10^{-7}	1.50×10^{-2}

In the case of the hydroxides, ΔH_{latt} is the more important factor. This is because the hydroxide ion is quite small, so the sum of the radii of the cation and the anion is significantly influenced by the cation size. This is seen clearly in the smaller differences in solubility for the hydroxides of the largest cations in the group (Table 1.5). The lattice enthalpies of the hydroxides decrease as the cation gets larger since the cation–anion distance is not dominated by a large anion. Thus ΔH_{latt} decreases more rapidly than the enthalpies of hydration.

Questions

1 Use the following data/kJ mol^{-1} to find the lattice energy of lithium fluoride LiF:
first ionisation energy of lithium = +520, enthalpy of atomisation of lithium = +159, enthalpy of atomisation of fluorine = +79, electron affinity of fluorine atoms = −328, enthalpy of formation of LiF = −616

2 (a) Find the lattice energy for calcium sulphide CaS from a Born – Haber cycle, given the following data/kJ mol^{-1}:
first ionisation energy of Ca = +590; second ionisation energy of Ca = +1145; first electron affinity of sulphur = −200; second electron affinity of sulphur + +640; enthalpy of atomisation of sulphur = +278; enthalpy of atomisation of calcium +178; enthalpy of formation of CaS = −482.

   (b) Suggest reasons for the difference between the value obtained and the lattice energy for CaO, which is −3401 kJ mol^{-1}.

3 Use the following data/kJ mol^{-1} to find the enthalpy of solution of sodium chloride NaCl:
Lattice energy of sodium chloride = −771; hydration enthalpy of Na+ = −406; hydration enthalpy of Cl− = −364.

4 State and explain the trend in solubility for
   (a) the hydroxides of Group 2, Mg to Ba;
   (b) the sulphates of Group 2.

The Periodic Table II

Fig. 2.1 Dimitri Ivanovitch Mendeleev (1834–1907). Russian chemist who presented the basis of the Periodic Table in February 1869, predicting the existence of elements not then known.

The Mendeleev table

The Periodic Table is a great unifying idea in chemistry. It can tell you a lot about the chemistry of an element, and when it first appeared it fulfilled one of the greatest requirements of any discovery: not only did it bring together things already known, but it also enabled predictions, in this case about elements then unknown.

D.I. Mendeleev (Figure 2.1) was not the first to attempt an arrangement of the elements, but when he published his table in 1869 he was clever (or daft) enough firstly to order the elements by atomic 'weight' (i.e. atomic mass), but was prepared to ignore the result if it clearly didn't work in terms of the known chemistry; and secondly left gaps for elements yet to be discovered.

Mendeleev didn't have atomic numbers, the particles in the atom and its structure not being completely discovered for another 70 years or so. The use of relative atomic masses would have given a table with some anomalies, such as iodine being in Group 6 and tellurium in Group 7, which is absurd in chemical terms. Mendeleev ignored the masses and stuck by the chemistry. Only after the work of Bohr and his colleagues on electronic structure was it clear that electronic structures determine the chemistry of an element, and that the atomic number orders the table. The 'anomaly' arose because relative atomic mass does not always increase with an increase of atomic number. The gaps were a triumph for the predictive powers of the table. Mendeleev predicted ten elements, eight of which are now known. One does not exist since there is no gap in modern tables, and the other has a very high mass and the table does not (yet) go that far.

The arrangement by atomic number brings the elements into groups and periods, and the long form of the table is shown in the Appendix.

Aspects of the chemistry of Period 3: sodium to chlorine

The ideas of electronegativity, size and bond type of atoms which were covered in Unit 1 are illustrated well by the elements of Period 3. The reactions which are used to illustrate trends can be used for any period in the Periodic Table.

Reactions of the elements

The reactions with oxygen, water and chlorine, summarised in Tables 2.1 – 2.3 are of most significance because oxygen and water are plentiful in the atmosphere, and chlorine is a typical strong oxidising agent.

The electropositive metals (Na and Mg) with low ionisation energies and fairly large ions give ionic products. Aluminium has a small ion with high charge and is quite polarising, so the oxide is ionic but the chloride is covalent since the chloride ion is larger and so more polarisable than the oxide ion. The non-metals react with varying vigour to give covalent products as would be expected from their high ionisation energies and small size.

Table 2.1 *Reactions of the elements of Period 3 with oxygen*

Na	Tarnishes in air; burns with yellow flame on heating to give a mixture of the oxide and peroxide, a yellowish-white solid. $4Na(s) + O_2(g) \rightarrow 2Na_2O(s)$ and $2Na(s) + O_2(g) \rightarrow Na_2O_2(s)$
Mg	Superficial oxidation at room temperature; burns with brilliant white flame on heating. Oxide white, ionic. $2Mg(s) + O_2(g) \rightarrow 2MgO(s)$
Al	Superficial oxide layer forms at room temperature, which protects against further attack except by concentrated alkali, and enables the use of this relatively reactive metal for engineering purposes. Burns on heating in oxygen to give white, ionic oxide $4Al(s) + 3O_2(g) \rightarrow 2Al_2O_3(s)$
Si	No reaction at room temperature. Burns on heating in oxygen to give white, giant covalent oxide. $Si(s) + O_2(g) \rightarrow SiO_2(s)$
P	White phosphorus catches fire spontaneously in air; red phosphorus burns on heating, to give the white, covalent oxide of phosphorus (+5). $P_4(s) + 5O_2(g) \rightarrow P_4O_{10}(s)$
S	Burns with bright blue flame on heating in air or oxygen, giving colourless covalent gaseous oxide. $S(s) + O_2(g) \rightarrow SO_2(g)$
Cl	Does not react directly with oxygen.

Table 2.2 *Reactions of the elements of Period 3 with chlorine*

Na	Reacts on heating to give white, ionic chloride. $2Na(s) + Cl_2(g) \rightarrow 2NaCl(s)$
Mg	Reacts on heating to give white, ionic chloride. $Mg(s) + Cl_2(g) \rightarrow MgCl_2(s)$
Al	Reacts on heating to give pale-yellow covalent chloride. $2Al(s) + 3Cl_2(g) \rightarrow 2AlCl_3(s)$
Si	Reacts on heating to give colourless, covalent liquid chloride. $Si(s) + 2Cl_2(g) \rightarrow SiCl_4(l)$
P	Reacts on heating to give colourless, covalent liquid trichloride. $2P(s) + 3Cl_2(g) \rightarrow 2PCl_3(l)$ Excess chlorine oxidises this product to a white solid containing PCl_4^+ and PCl_6^- ions; in the gas phase it consists of PCl_5 molecules. $PCl_3(l) + Cl_2(g) \rightarrow PCl_5(s)$. This reaction is an equilibrium in a closed system.
S	Reacts on heating to give a pale-yellow covalent liquid. $2S(s) + Cl_2(g) \rightarrow S_2Cl_2(l)$

THE PERIODIC TABLE II

QUESTION

What type of bonding would you expect in aluminium fluoride? Why?

Table 2.3 *Reactions of the elements of Period 3 with water*

Na	Violent reaction at room temperature; metal melts and rushes about on the surface of the water; hydrogen evolved, product solution alkaline. $2Na(s) + 2H_2O(l) \rightarrow 2NaOH(aq) + H_2(g)$
Mg	Extremely slow reaction with cold water. Steam passed over heated metal gives very exothermic reaction (metal glows red) to give the oxide. $Mg(s) + H_2O(g) \rightarrow MgO(s) + H_2(g)$
Al	No reaction with cold water. Steam passed over finely divided, heated metal gives the oxide. $2Al(s) + 3H_2O(l) \rightarrow Al_2O_3(s) + 3H_2(g)$
Si P S	} Do not react with water.
Cl	Forms acidic solution containing hydrated chlorine molecules as well as chloride and chlorate(I) ions from disproportionation. $Cl_2(aq) + H_2O(l) \rightleftharpoons HOCl(aq) + HCl(aq)$

QUESTION

Draw a dot-and-cross diagram for aluminium chloride dimer, Al_2Cl_6.

The oxides of Period 3

The formulae and bonding of the principal oxides, together with their acid/base properties, are given in Figure 2.2.

Na	Mg	Al	Si	P	S	Cl
Na_2O	MgO	Al_2O_3	SiO_2	P_4O_{10}	SO_2 SO_3	Cl_2O
ionic			giant covalent	molecular covalent		
basic		amphoteric	acidic			

DEFINITION

A basic oxide reacts with an acid to give a salt and water. An acidic oxide reacts with a base to give a salt and water. An amphoteric oxide will react with either an acid or a base to give a salt and water.

Fig. 2.2 The important oxides of Period 3.

The s-block metals have basic oxides, which react with acids to give salts:

$$Na_2O(s) + 2HCl(aq) \rightarrow 2NaCl(aq) + H_2O(l)$$

$$MgO(s) + 2HCl(aq) \rightarrow MgCl_2(aq) + H_2O(l)$$

The hydroxides are also basic:

$$NaOH(aq) + HCl(aq) \rightarrow NaCl(aq) + H_2O(l)$$

$$Mg(OH)_2(s) + 2HCl(aq) \rightarrow MgCl_2(aq) + 2H_2O(l)$$

Aluminium oxide, although ionic, is amphoteric, and will react with both acid and base:

$$Al_2O_3(s) + 6H^+(aq) \rightarrow 2Al^{3+}(aq) + 3H_2O(l)$$

$$Al_2O_3(s) + 3H_2O(l) + 6OH^-(aq) \rightarrow 2[Al(OH)_6]^{3-}(aq)$$

In practice these reactions are difficult to show unless the oxide is freshly prepared. Ageing for a few hours changes the oxide structure to one which is resistant to attack except by concentrated alkali. The amphoteric nature of aluminium oxide classifies aluminium as a semi-metal on chemical grounds, that is, an element whose chemistry shows the characteristics of both metallic and non-metallic elements. If classified on electrical conductivity and other physical properties, aluminium is a metal.

Aluminium hydroxide reacts readily with acid and base; addition of aqueous sodium hydroxide to a solution of aluminium ions gives a white gelatinous precipitate of aluminium hydroxide. This disappears with addition of excess alkali; if acid is added to the solution, the white precipitate of aluminium hydroxide returns, then with more acid gives a solution of the aluminium salt once more:

$$Al^{3+}(aq) \underset{3H^+}{\overset{3OH^-}{\rightleftharpoons}} Al(OH)_3(s) \underset{3H^+}{\overset{3OH^-}{\rightleftharpoons}} Al(OH)_6^{3-}(aq)$$

colourless solution   white precipitate   colourless solution

$3H_2O(l)$ $3H_2O(l)$

The non-metal oxides of Period 3 are all acidic. Silicon dioxide will react with molten sodium hydroxide:

$$SiO_2(s) + 2NaOH(l) \rightarrow Na_2SiO_3(l) + H_2O(g)$$

or with calcium oxide in a blast furnace,

$$SiO_2(s) + CaO(s) \rightarrow CaSiO_3(l)$$

It does not react with aqueous alkali except under pressure at high temperature.

Phosphorus(V) oxide, P_4O_{10}, reacts violently with water to give phosphoric(V) acid

$$P_4O_{10}(s) + 6H_2O(l) \rightarrow 4H_3PO_4(aq)$$

and has such an affinity for water that it makes an excellent, though expensive, drying agent, and will dehydrate concentrated sulphuric acid to sulphur trioxide.

Fig. 2.3 Blast furnace

Sulphur dioxide gives sulphurous acid (sulphuric(IV) acid) in water

$$SO_2(g) + H_2O(l) \rightarrow H_2SO_3(aq)$$

although the acid exists only in dilute solution. Sulphur trioxide reacts so violently with water to give sulphuric acid (sulphuric(VI) acid) that in the manufacture of this important material the sulphur trioxide is dissolved in 98% sulphuric acid forming acid of 98.5% concentration. This is then diluted with water back to its original concentration. A mist of sulphuric acid is otherwise formed which is very difficult to handle.

Dichlorine oxide is a brownish-yellow gas that reacts with water to give a solution of chloric(I) acid (hypochlorous acid):

$$Cl_2O(g) + H_2O(l) \rightarrow 2HOCl(aq)$$

The acid/base nature of oxides thus depends on the character of the element with which the oxygen is combined. The more electropositive this element is, the more basic the oxide, since there is no (or little) covalent bonding and the oxide ion will readily accept hydrogen ions to form water. With the covalent oxides, the bonds with the other element are strong although polar, and there is no oxide ion to receive protons. This results in the reaction with the polar water forming products where covalent bonding between the original element and oxygen survives. For sulphur dioxide in water, therefore:

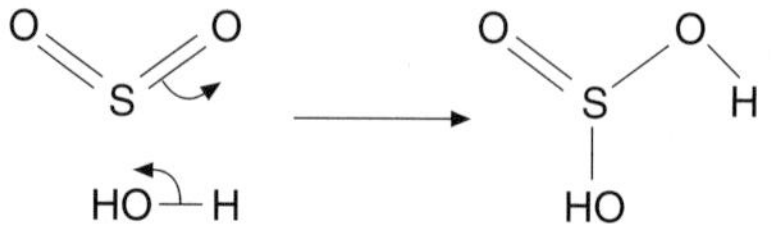

The chlorides of Period 3

The formulae and bonding of the principal chlorides of Period 3 elements are given in Figure 2.4. The formulae of the chlorides show the expected progression based on the group oxidation state of the elements.

Phosphorus and sulphur in higher oxidation states form other chlorides, e.g. PCl_5, SCl_4.

The ionic chlorides dissolve in water to give the hydrated ions and neutral solutions. The covalent chlorides, though, have polar bonds which undergo nucleophilic attack by the lone pair on the oxygen of the water molecule. The reactions are violent apart from that of chlorine; and all apart from chlorine fume in moist air owing to the production of hydrogen chloride droplets.

QUESTION

Look up in a data book the melting and boiling temperatures and densities of aluminium and of phosphorus to compare the physical properties of these Period 3 elements (metal and non-metal.)

Na	Mg	Al	Si	P	S	Cl
NaCl	MgCl$_2$	AlCl$_3$	SiCl$_4$	PCl$_3$ PCl$_5$	S$_2$Cl$_2$	Cl$_2$
ionic		covalent		molecular covalent		

Fig. 2.4 Chlorides of Period 3 elements.

Aluminium chloride at room temperature has a layer lattice with a high degree of covalent character. At higher temperatures, it starts to form Al_2Cl_6 dimers, and just below its melting temperature the volume suddenly increases by some 85% as the dimers form in large numbers.

The hydrolysis of aluminium chloride with a *small amount* of water proceeds in stages:

$$AlCl_3(s) + H_2O(l) \rightleftharpoons AlCl_2OH(aq) + HCl(aq)$$

$$AlCl_2OH(aq) + H_2O(l) \rightleftharpoons AlCl(OH)_2(aq) + HCl(aq)$$

$$AlCl(OH)_2(aq) + H_2O(l) \rightleftharpoons Al(OH)_3(s) + HCl(aq)$$

Aluminium chloride produces an acidic solution in a large amount of water but no precipitate of aluminium hydroxide.

The acidity is due to the interaction of the water with the hexa-aqua-aluminium ion:

$$[Al(H_2O)_6]^{3+}(aq) + H_2O(l) \rightarrow [Al(H_2O)_5(OH)]^{2+} + H_3O^+(aq)$$

This deprotonation reaction does not proceed much further unless sodium hydroxide is added, in which case $Al(OH)_3$ will be precipitated.

In the case of silicon tetrachloride, although in this case the hydrolysis is complete. The product, silica, is initially formed as a hydrate which loses water to give SiO_2.

$$SiCl_4(l) + 2H_2O(l) \rightarrow SiO_2(s) + 4HCl(aq)$$

The hydrolysis of the remaining chlorides apart from S_2Cl_2 is represented by:

$$PCl_3(l) + 3H_2O(l) \rightarrow H_3PO_3(aq) + 3HCl(aq)$$

$$PCl_5(s) + 4H_2O(l) \rightarrow H_3PO_4(aq) + 5HCl(aq)$$

$$Cl_2(aq) + H_2O(l) \rightleftharpoons HOCl(aq) + HCl(aq)$$

The hydrolysis of disulphur dichloride is complex and cannot be represented by a single equation. The acidic solution of the hydrolysis products includes sulphur, sulphite ions, thiosulphate ions, hydrogen sulphide, and a range of thionic acids $H_2S_nO_6$, where n is from 2 to 6 or more.

The increase in metallic character in Group 4

The chemistry of the elements in Group 4 changes from that of a non-metal in carbon to that of a metal in lead. Metallic character, chemically speaking, is considered to involve:

- the formation of positive ions
- the formation of predominantly ionic chlorides which do not hydrolyse significantly when placed in water
- the formation of ionic oxides which are basic or amphoteric.

The physical properties of the Group 4 elements change from those of typical non-metals to those of metals (Table 2.4). Don't be fooled by graphite's conductivity – it is not the same as that of metals, because metals conduct equally in *all* directions, whereas graphite only conducts parallel to the layers in the crystal.

Carbon as a non-metal

Carbon forms covalent bonds; the isolated C^{4+} ion does not exist. This can be explained in two ways: C^{4+} would have an estimated ionic radius of 15 pm. Such a tiny, highly charged ion would be enormously polarising and would form covalent bonds. Alternatively: **ionic compounds are favoured if the ionisation energies of the cation (endothermic) and the electron affinities of the anion (which overall may be exo- or endothermic) are compensated by the exothermic lattice enthalpy of the resulting compound**. The first four ionisation energies of carbon are in total 14 270 kJ mol^{-1}, and there is no chance of recouping this through the lattice energy of any conceivable ionic solid.

Tetrachloromethane CCl_4 is liquid at room temperature, and is molecular covalent. It does not hydrolyse with water, unlike many covalent chlorides, but this is a kinetic effect which is explained below (page 21).

Carbon dioxide is acidic; it is a typical non-metal oxide (see Table 2.5 later).

Lead as a metal

Lead is really a semi-metal, showing characteristics of both metal and non-metal in its chemistry. The metallic nature predominates, however, and physical properties (density, conductivity, lustre) are metallic (Table 2.4).

Table 2.4 *Physical properties of the Group 4 elements*

Element	Appearance	Bonding	Density/g cm^{-3}	Melting temperature/°C	Boiling temperature/°C
carbon	shiny black solid (graphite)	giant covalent layer lattice	2.25	3652 (sublimes)	4827
	colourless solid (diamond)	giant covalent	3.51	>3550	4827
silicon	shiny dark grey-blue solid	giant covalent	2.32	1410	2355
tin	yellowy-silver solid	metallic	7.28	232	2270
lead	greyish-silver solid	metallic	11.34	328	1740

The density shows a gradation from typically non-metallic low values to the higher density associated with metals. The melting temperatures, though, are not typically non-metallic for carbon and silicon; instead they are a reflection of the fact that these atoms can form four covalent bonds and therefore are able to produce very large three-dimensional giant molecules.

As the size of the atoms in the group increases, the ionisation energies fall. With lead (+2) the first two ionisation energies are sufficiently low to result in $PbCl_2$ being ionic and not hydrolysed by water. PbO and PbO_2 are both amphoteric; PbO_2 in addition is strongly oxidising.

The oxides of Group 4

Table 2.5 gives a summary of the main properties of these oxides. As expected, the trend from non-metal to metal in both groups is shown by increasing basicity and ionic character in the oxides.

Carbon

The acidity of carbon dioxide is shown in its reaction with water. This gives a solution of carbonic acid which, when saturated with the gas, is around pH 5:

$$CO_2(aq) + H_2O(l) \; \rightleftharpoons \; H^+(aq) + HCO_3^-(aq)$$

$$HCO_3^-(aq) \; \rightleftharpoons \; H^+(aq) + CO_3^{2-}(aq)$$

Carbonic acid forms two series of salts, the hydrogen carbonates (bicarbonates) such as $NaHCO_3$, and the carbonates, formed by nearly all metals. Aluminium and beryllium are notable exceptions.

Table 2.5 *Some properties of the oxides of Group 4*

Oxide	Nature	Example reactions
CO	Neutral	Does not react with water. With hot NaOH solution under pressure CO gives sodium methanoate: $CO + NaOH \rightarrow HCOONa$.
CO_2	Acidic	Dissolves in water to give the weak acid carbonic acid, H_2CO_3. $H_2CO_3 + H_2O \rightleftharpoons H_3O^+ + HCO_3^-$ $HCO_3^- + H_2O \rightleftharpoons CO_3^{2-} + H_3O^+$
SiO_2	Acidic	$SiO_2 + CaO \rightarrow CaSiO_3$ in blast furnace. NaOH needs to be molten or in hot concentrated aqueous solution: $SiO_2 + 2NaOH \rightarrow Na_2SiO_3 + H_2O$
SnO	Amphoteric	Basic nature: $SnO + 2HCl \rightarrow SnCl_2 + H_2O$ Acidic nature: $SnO + 2NaOH + H_2O \rightarrow Na_2Sn(OH)_4$
SnO_2	Amphoteric	Basic nature: $SnO_2 + 4HCl \rightarrow SnCl_4 + 2H_2O$ Acidic nature: $SnO_2 + 2NaOH \rightarrow Na_2SnO_3 + H_2O$
PbO	Amphoteric	Basic nature: $PbO + 2HCl \rightarrow PbCl_2 + H_2O$. $PbCl_2$ dissolves in conc HCl to give soluble ions such as $PbCl_3^-$ and $PbCl_4^{2-}$. Acidic nature: $PbO + 2NaOH + H_2O \rightarrow Na_2Pb(OH)_4$, sodium plumbate(II).
PbO_2	Amphoteric	Basic nature: with ice-cold conc HCl, $PbCl_4$ is formed. If this is allowed to warm up it decomposes to chlorine and $PbCl_2$: this shows the (much more important) oxidising nature of lead(IV) oxide: $PbO_2 + 4HCl \rightarrow PbCl_4 + 2H_2O$ $PbCl_4 \rightarrow PbCl_2 + Cl_2$ Most other acids do not affect PbO_2. Acidic nature: with molten NaOH: $PbO_2 + 2NaOH \rightarrow Na_2PbO_3$ (sodium plumbate(IV)) $+ H_2O$.
Pb_3O_4	Behaves as a mixture of PbO_2 and PbO	$Pb_3O_4 + 4HNO_3 \rightarrow 2Pb(NO_3)_2 + PbO_2 + 2H_2O$ This shows the greater acidity of PbO_2 compared with PbO. Lead(IV) oxide is not basic enough to react with dilute nitric acid.

Carbon monoxide is, in principle, the anhydride of methanoic acid, and might be expected to react with water according to the equation:

$$CO + H_2O \rightarrow HCOOH$$

Carbon monoxide can be made by the dehydration of methanoic acid with concentrated sulphuric acid. However, carbon monoxide is inert to water, and the above reaction does not occur. It will react with aqueous sodium hydroxide under pressure at high temperature to give sodium methanoate:

$$CO(g) + NaOH(aq) \rightarrow HCOONa(aq)$$

The acidic properties of carbon monoxide are so weak that it is regarded as a neutral oxide.

Silicon

Silicon(IV) oxide, silica, is acidic, but it is so insoluble in water that it does not noticeably react with aqueous alkali. In the purification of bauxite for aluminium production, silica does not interfere with the process since the 10% aqueous solution of sodium hydroxide used in current practice does not dissolve the silica impurity in the bauxite. Silica will react with molten sodium hydroxide to give sodium silicate, Na_2SiO_3, which is an ingredient of detergents for dishwashers. In the blast furnace, the reaction with calcium oxide is important in removing SiO_2 impurity from the iron ore and enabling a continuous process to occur.

Tin

Both tin(II) oxide, SnO, and tin(IV) oxide SnO_2, are amphoteric oxides, and react with hydrochloric acid in acid/base reactions to give the corresponding chloride:

$$SnO(s) \ + \ 2HCl(aq) \ \rightarrow \ SnCl_2(aq) \ + \ H_2O(l)$$

$$SnO_2(s) \ + \ 4HCl(aq) \ \rightarrow \ SnCl_4(aq) \ + \ 2H_2O(l)$$

Tin(IV) chloride is covalent, so the solution has to be kept acidic if the following hydrolysis is to be avoided:

$$SnCl_4(aq) \ + \ H_2O(l) \ \rightleftharpoons \ SnCl_3OH(aq) \ + \ HCl(aq)$$

With sodium hydroxide these oxides give soluble salts, sodium stannate(II) and sodium stannate(IV). They are not important.

Lead

Lead(IV) oxide is amphoteric. However, it is resistant to attack by most acids other than concentrated hydrochloric acid, and reacts only with molten, rather than aqueous, sodium hydroxide. The acid/base properties of lead(IV) oxide are unimportant. Much more significant is its oxidising power.

The amphoteric nature of lead(II) oxide is shown by the following reactions:

As a base, the reaction

$$PbO(s) + 2HCl(aq) \rightarrow PbCl_2(s) + H_2O(l)$$

occurs. In practice, if concentrated hydrochloric acid, is used, solutions of complexes such as $PbCl_3^-$ and $PbCl_4^{2-}$ are obtained. These precipitate white lead(II) chloride, $PbCl_2$, on dilution. The reaction of lead(II) oxide with sulphuric acid stops almost immediately since lead sulphate, $PbSO_4$, is insoluble. With alkali, lead(II) oxide reacts as an acid and the soluble plumbate(II) is formed:

$$PbO(s) + 2OH^-(aq) + H_2O(l) \rightarrow Pb(OH)_4^{2-}(aq)$$

The chlorides of Group 4

The principal features of Group 4 chlorides are summarised in Table 2.6 together with the bond type and the reaction with water. Germanium chlorides are unimportant.

Table 2.6 *Some properties of the Group 4 chlorides*

	Chloride	Nature and reaction with water	Chloride	Nature and reaction with water
Carbon			CCl_4	Covalent liquid, unreactive below 1000°C
Silicon			$SiCl_4$	Covalent liquid, hydrolyses
Germanium	$GeCl_2$	Ionic, hydrolyses	$GeCl_4$	Covalent liquid, hydrolyses
Tin	$SnCl_2$	Ionic, hydrolyses	$SnCl_4$	Covalent liquid, hydrolyses
Lead	$PbCl_2$	Ionic, insoluble	$PbCl_4$	Covalent liquid, unstable above 0°C, hydrolyses

The larger atoms in Group 4 give halides in oxidation state +2 which hydrolyse in water, though not completely. They form basic chlorides:

$$SnCl_2(s) + H_2O(l) \rightleftharpoons SnClOH(s) + HCl(aq)$$

The cloudy liquid which results can be cleared by the addition of concentrated hydrochloric acid, which shifts this equilibrium to the left-hand side. Solutions of tin(II) chloride must, therefore, be acidic if hydrolysis is to be avoided. The tetrahalides show marked differences in properties. Tetrachloromethane, CCl_4, is not attacked by water at room temperature, and even at 1000°C it hydrolyses only partially:

$$CCl_4(g) + H_2O(g) \rightarrow COCl_2(g) + 2HCl(g)$$

The product, carbonyl chloride, is extremely toxic. The resistance of tetrachloromethane to hydrolytic attack is dealt with below (page 23).

Oxidation states in Group 4

Carbon shows a range of oxidation states: +4 in CO_2, +2 in CO and –4 in CH_4. Tin and lead show +4 and +2, and +2 becomes more stable as the group is descended, in practice the difference being marked between tin and lead. Thus tin (+2) is reducing and finds a use in the production of phenylamine from nitrobenzene:

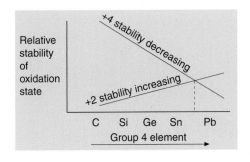

NO_2 ... $+ 3Sn^{2+} + 7H^+ \rightarrow$ ... $\overset{+}{N}H_3$ ... $+ 2H_2O + 3Sn^{4+}$

A mixture of tin and concentrated hydrochloric acid is used, which react when heated under reflux to give tin(II) chloride which is the reducing agent. Lead (+2) shows no such reducing properties.

Tin(IV) oxide does not oxidise chloride ions; with lead(IV) oxide, PbO_2, the situation is wholly different. The +4 state of lead is strongly oxidising, and at room temperature concentrated HCl is oxidised to chlorine, lead (+4) being reduced to lead (+2).

$$PbO_2(s) \quad + \quad 4HCl(aq) \quad \rightarrow \quad PbCl_2(s) \quad + \quad Cl_2(g) \quad + \quad 2H_2O(l)$$

Lead (+2) shows no reducing properties.

BACKGROUND

The reason for these differences is related to the sizes of the atoms. The electron structure of Group 4 atoms is $ns^2\,np^2$. The atom could form an ion by loss of the p electrons, or two covalent bonds by sharing, again using the p-orbitals. Both would give compounds with the Group 4 atom in the +2 oxidation state. The (+4) state could be achieved by loss of all four electrons to give a 4+ ion, but even in the case of lead such an ion would be small enough to be very polarising, and the resultant compounds would be covalent. Alternatively, if one of the s electrons is promoted to the p, with one electron in each orbital, such an excited state can form four covalent bonds:

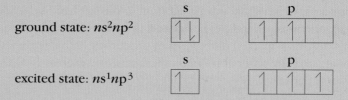

Fig. 2.5 *Formation of covalent bonds by electron promotion.*

The electron promotion is endothermic. It will occur if the energy required can be repaid by formation of the extra bonds, that is if the E–X (where E=element) bond strength is about a quarter of the promotional energy or more. This in turn will depend on the bond length: Table 2.7 gives some values for bonds with chlorine. The bond strengths have been calculated from Born–Haber cycles.

Table 2.7 *Bond lengths and bond strengths between Group 4 elements and chlorine*

Bond	Bond length/pm	Bond strength/kJ mol^{-1}
C–Cl	177	346
Si–Cl	203	407
Sn–Cl	220	324
Pb–Cl	253	252

In the case of tin, the bonds formed are strong enough to compensate for the promotional energy of the s electron, so tin(IV) chloride is more thermodynamically favoured than tin(II) chloride. With lead, the longer, weaker Pb–Cl bonds do not compensate energetically, and so ionic $PbCl_2$ with a lattice enthalpy of -2269 kJ mol^{-1} is thermodynamically more stable than covalent $PbCl_4$.

The hydrolysis of $SiCl_4$ and CCl_4

Tetrachloromethane CCl_4 and silicon tetrachloride $SiCl_4$ show marked differences from one another in their behaviour towards water. Both are molecular covalent; the bonds are polar, but the tetrahedral shape of the molecule means that the dipoles cancel and the molecules have no overall dipole moment.

Tetrachloromethane is not attacked by water at room temperature; the liquids are immiscible. Even at 1000°C it hydrolyses only partially:

$$CCl_4 + H_2O \rightarrow COCl_2 + 2HCl$$

The carbonyl chloride produced (phosgene) is extremely toxic and has been used as a poison gas in warfare. Tetrachloromethane was at one time used in portable fire extinguishers for electrical fires, but carbonyl chloride was produced when these were used. They are obsolete!

The resistance of tetrachloromethane to hydrolysis is due to the high activation energy for the reaction. Thermodynamically, the hydrolysis is favourable. Using enthalpies of formation ΔH_f, we can find ΔH for the reaction

$$CCl_4 + 2H_2O \rightarrow CO_2 + 4HCl$$

$$\Delta H = \Delta H_f^{\ominus}(CO_2) + 4\Delta H_f^{\ominus}(HCl) - \Delta H_f^{\ominus}(CCl_4) - 2\Delta H_f^{\ominus}(H_2O)$$

$$= (-393.5) + 4(-92.3) - (-129.6) - 2(-285.8)$$

$$= -61.5 \text{ kJ mol}^{-1}$$

The mixture of tetrachloromethane and water is thermodynamically unstable. Tetrachloromethane has no energetically accessible vacant orbitals (i.e. there is no $2d$) that could accept a lone electron pair from the oxygen of the attacking water molecule since all are used in bonding. In order for the reaction to occur, C–Cl bonds must be broken or at least weakened, and this results in a large activation energy for the process. The mixture is therefore kinetically stable.

In addition, there is a steric effect. The chlorine atoms around the carbon are quite bulky, and prevent access of the attacking water molecule to the carbon atom. This is called **steric hindrance**. The chlorine atoms are so bulky that

QUESTION

What is the difference between **thermodynamic** stability and **kinetic** stability?

they repel one another in CCl_4, and this is the reason why the C–Cl bond *in this compound* is weaker than the Si–Cl bond in $SiCl_4$.

This kinetic stability is a highly significant feature of carbon chemistry generally. Most carbon compounds are thermodynamically unstable with respect to oxidation or to hydrolysis – yet life needs oxygen and water. Fortunately the compounds are kinetically stable. Most reactions in inorganic chemistry operate under thermodynamic, not kinetic, control.

Silicon tetrachloride, $SiCl_4$, is rapidly attacked by water at room temperature:

$$SiCl_4(l) \; + \; 2H_2O(l) \; \rightarrow \; SiO_2(s) \; + \; 4HCl(aq)$$

The initial product is hydrated silica, $Si(OH)_4$, but this loses water quickly. The rapidity of the hydrolysis is due to the electron-pair donation by the oxygen of water into empty 3d orbitals of the silicon atom. In this hydrolysis Si–Cl bonds do not have to be broken before Si–O bonds start to form. The activation energy for this hydrolysis is therefore much lower than that for tetrachloromethane. The steric effects in the larger molecule are also much less important.

Questions

1   Plot small graphs of (a) atomic radius, (b) first ionisation energy, for the elements of Period 3, Na to Cl. Comment on any relationship between these graphs.

2   Look up the melting temperatures of silicon, phosphorus, sulphur and chlorine, and find out why they show the pattern that they do.

3   A scarlet oxide of lead, **A**, which contains the metal in two different oxidation states, reacts with aqueous nitric acid to give a solution of compound **B**, and a brown residue **C**. With sodium hydroxide, solution **B** gives a white precipitate **D** which, with further addition of sodium hydroxide, gives a colourless solution **E**.

Residue **C** reacts with concentrated hydrochloric acid to give a pale-yellow solution **F** and a green gas **G**, which will oxidise iodide ions to iodine. If **F** is diluted with water, a white solid **H** is precipitated.
Identify all the substances **A** to **H**, and give equations to represent all the reactions occurring.

Chemical equilibrium II

The equilibrium constants, K_c and K_p

The position of equilibrium at a given temperature can be defined by the **equilibrium constant**, which is given the symbol K_c, or K_p for gaseous systems.

The equilibrium constant K_c

This constant relates the equilibrium concentrations, measured in $mol\,dm^{-3}$, of the substances present in the equilibrium mixture. Thus for the general case of a homogeneous equilibrium involving substances A, B, C, P and Q, for which the equation is:

$$n\text{A} + m\text{B} + x\text{C} \rightleftharpoons p\text{P} + y\text{Q}$$

the equilibrium constant K_c is given by the expression:

$$K_c = \frac{[\text{P}]^p[\text{Q}]^y}{[\text{A}]^n[\text{B}]^m[\text{C}]^x}$$

It is important to note that:
- the symbol [ ] represents the mole concentration of a particular species **at equilibrium** in $mol\,dm^{-3}$ and should, strictly speaking, be written $[\ ]_{Equ}$.
- by convention the substances on the right-hand side of the equation are placed in the numerator of the expression
- the concentration of each substance is raised to a power which is the same as the number of moles of that substance in the equation
- all substances present in the equilibrium mixture must be included in the expression for K_c *unless* they are solids
- K_c must be quoted together with the equation for the reaction.

The value of the equilibrium constant K_c depends *only* on the temperature at which the equilibrium is established. It does not depend on the concentrations of the substances used initially, nor on the equilibrium concentrations which are related to each other and must be such that they maintain K_c constant at a given temperature.

For example, in the reaction:

$$2SO_2(g) + O_2(g) \rightleftharpoons 2SO_3(g)$$

the expression for the equilibrium constant K_c will be:

$$K_c = \frac{[SO_3]^2}{[SO_2]^2[O_2]}$$

The units of K_c

The units of K_c will depend on the actual expression for K_c. For example, in the expression for the sulphur dioxide/oxygen/sulphur trioxide equilibrium above, the units for K_c would be:

$$K_c = \frac{(\text{mol dm}^{-3})^2}{(\text{mol dm}^{-3})^2 \times (\text{mol dm}^{-3})} = (\text{mol dm}^{-3})^{-1} = \text{mol}^{-1}\,\text{dm}^3$$

In the reaction described below, K_c has no units.

Experimental determination of K_c values

In order to determine the values of K_c by experiment, it is necessary to set up an equilibrium at a given temperature, determine the concentrations of each constituent of the equilibrium mixture, then substitute these values into the expression for the equilibrium constant. As it requires time to determine these concentrations, it is necessary to 'freeze' the equilibrium so that the position of equilibrium does not change while the concentrations are being determined. This is usually done by rapidly reducing the temperature to a level where reaction ceases. One reaction which is relatively easy to study in the school laboratory is that between a carboxylic acid and an alcohol, for example:

$$CH_3COOH(l) + C_2H_5OH(l) \rightleftharpoons CH_3COOC_2H_5(l) + H_2O(l)$$

Mixtures of acid and alcohol are heated under reflux, in a thermostatically controlled bath at say 60°C, for several hours in order to achieve equilibrium. The mixture is rapidly cooled by placing it in a large bath of cold water containing ice. The amount of carboxylic acid remaining can then be determined by titration with standard sodium hydroxide using a suitable indicator such as phenolphthalein. The concentrations of the other substances in the equilibrium mixture need not be experimentally determined since they can be deduced from the stoichiometry of the reaction.

The equilibrium could also be approached from a mixture of ethyl ethanoate and water.

Equilibrium mixture

Fig. 3.1 An equilibrium can be approached from either end.

For example, 1 mol of ethanoic acid and 1 mol of ethanol were mixed and heated under reflux at 333 K for several hours until equilibrium was established. The contents of the flask were then poured into some cold water in order to stop any further reaction, and then made up to 1.00 dm³ with distilled water. Portions of this solution, each of 25.0 cm³, required 27.5 cm³ of 0.300 mol dm⁻³ sodium hydroxide for complete neutralisation using phenolphthalein as indicator.

Thus the amount of NaOH required = $0.0275 \text{ dm}^3 \times 0.300 \text{ mol dm}^{-3}$

$$= 8.25 \times 10^{-3} \text{ mol}$$

The equation for the reaction between NaOH and ethanoic acid is:

$$CH_3COOH(aq) + NaOH(aq) \rightarrow CH_3COO^-Na^+(aq) + H_2O(l)$$

Hence the amount of ethanoic acid remaining in $25.0\,cm^3$ of the equilibrium mixture is 8.25×10^{-3} mol.

The number of moles of ethanoic acid in $1\,dm^3$ of solution
$$= 40 \times 8.25 \times 10^{-3}\,mol$$

$$= 0.33\,mol$$

This is therefore the equilibrium concentration of ethanoic acid. Since 1 mol of acid was present initially, 0.67 mol must have been converted into ester and water. Thus, using the stoichiometry of the equilibrium reaction:

$$CH_3COOH(l) + C_2H_5OH(l) \rightleftharpoons CH_3COOC_2H_5(l) + H_2O(l)$$

the concentrations of ethyl ethanoate and water at equilibrium must each be $0.67\,mol\,dm^{-3}$.

Also, the concentration of ethanol in the equilibrium mixture can be deduced since it reacts with the ethanoic acid in a 1:1 ratio, hence its concentration is also $0.33\,mol\,dm^{-3}$.

All this information can be summarised as shown below:

$$CH_3COOH(l) + C_2H_5OH(l) \rightleftharpoons CH_3COOC_2H_5(l) + H_2O(l)$$

Initial concentrations $(mol\,dm^{-3})$: 1.00	1.00	0	0
Equilibrium concentrations $(mol\,dm^{-3})$: 0.33	0.33	0.67	0.67

The equilibrium constant can then be calculated:

$$K_c = \frac{0.67\,mol\,dm^{-3} \times 0.67\,mol\,dm^{-3}}{0.33\,mol\,dm^{-3} \times 0.33\,mol\,dm^{-3}} = 4.12$$

K_c has no units in this case.

Other experiments done for the same equilibrium at the same temperature might give the results shown in Table 3.1.

Table 3.1 *Equilibrium concentrations on varying the concentration of reactants*

Initial concentrations /mol dm⁻³		Equilibrium concentrations /mol dm⁻³			
CH_3CO_2H	C_2H_5OH	CH_3CO_2H	C_2H_5OH	$CH_3CO_2C_2H_5$	H_2O
1.00	1.00	0.33	0.33	0.67	0.67
1.00	2.00	0.15	1.15	0.85	0.85
2.00	2.00	0.67	0.67	1.33	1.33

The first set of results are those found in the example above and lead to a value of K_c of 4.12. It would be instructive to verify that the other results lead to values of 4.19 and 3.94, respectively, which are in good agreement. Thus whatever the initial quantities, the concentrations of all components present at equilibrium will be such as to maintain a constant value for the equilibrium constant, provided that the temperature is constant.

Gaseous equilibria and K_p

Many equilibrium reactions exist which involve only gases. In such systems the concentration of the gases in moles per cubic decimetre can be used to find a value for the equilibrium constant K_c (see Unit 2). Nevertheless it is more convenient to measure the concentrations of gases in a mixture using the partial pressure of each of the gases.

Partial pressure

n moles of oxygen	n moles of hydrogen	$2n$ moles of hydrogen
Pressure P	The pressure is also P	The pressure is now $2P$

When a gas occupies a container of fixed volume at a given temperature, the pressure which is exerted by the gas depends only on the number of moles of gas present. The nature of the gas is not important. Under these conditions:

pressure exerted by the gas $\propto$ number of moles of the gas.

Thus:

$2n$ moles of oxygen	n moles of O_2 + n moles N_2
Pressure $2P$	The pressure is also $2P$

DEFINITION

Partial pressure (p):
the pressure a given gas in a gas mixture would exert if it alone filled the container at the same temperature. It is equal to the mole fraction of the gas multiplied by the total pressure.

The gas pressure P can be measured in any convenient units; atmospheres (atm) are commonly used in chemistry, but the use of kN m^{-2} (kPa) will also be encountered in examination questions.

Since the pressure of the gas is independent of the nature of the gas, then a mixture of gases will also exert a pressure which depends only on the number of moles of gas.

DEFINITION

Dalton's Law:
the sum of the partial pressure of each gas in a mixture equals the total pressure in the container.

In such a mixture of gases, each gas exerts a **partial pressure** p. This is the pressure that the gas would exert if it alone filled the container, at a given temperature. The sum of the partial pressures of each gas in a mixture equals the total pressure in the container; this is Dalton's Law. Thus in the mixture of oxygen and nitrogen, the oxygen exerts a pressure P and the nitrogen also exerts a pressure P.

The principle that applies to any gas mixture uses the **mole fraction** of the gas, x. The mole fraction of a gas G, x_G, is defined:

$$x_G \quad = \quad \frac{\text{number of moles of gas G}}{\text{total number of moles of gas in the system}}$$

If the mole fraction of each of the gases is known, together with the total pressure P_t, the partial pressure of each of the gases can be found. Each gas exerts a pressure which is this same fraction of the total pressure:

$$\text{Partial pressure of a gas} \quad = \quad \text{mole fraction} \quad \times \quad \text{total pressure}$$

$$p_G \quad = \quad x_G \times P_t$$

Partial pressures are therefore proportional to the number of moles of gas and are used as a measure of concentration in gas mixtures.

Example 1

In the equimolar mixture of nitrogen and oxygen considered initially:

$$\text{mole fraction of oxygen} \quad = \quad \frac{\text{number of moles of } O_2}{\text{number of moles of } O_2 + N_2}$$

$$= \quad \frac{n}{n+n}$$

$$= \quad \tfrac{1}{2}$$

The mole fraction of nitrogen is clearly the same. Thus the partial pressures of oxygen and nitrogen are both given by:

$$p_{O_2} \quad = \quad p_{N_2} \quad = \quad \tfrac{1}{2} \times 2P \quad = \quad P$$

Example 2

Now consider a mixture of n_{N_2} moles of nitrogen, n_{O_2} moles of oxygen and n_{CO_2} moles of carbon dioxide, in a container at a total pressure P_t. The total number of moles of gas present is:

$$n_{N_2} \quad + \quad n_{O_2} \quad + \quad n_{CO_2}$$

$$x_{N_2} = \frac{n_{N_2}}{n_{N_2} + n_{O_2} + n_{CO_2}} \quad \text{and}$$

$$x_{O_2} = \frac{n_{O_2}}{n_{N_2} + n_{O_2} + n_{CO_2}} \quad \text{and}$$

$$x_{CO_2} = \frac{n_{CO_2}}{n_{N_2} + n_{O_2} + n_{CO_2}}$$

and

$$p_{N_2} = x_{N_2} P_t \qquad \text{and} \qquad p_{O_2} = x_{O_2} P_t \qquad \text{and} \qquad p_{CO_2} = x_{CO_2} P_t$$

Suppose that a vessel contains 1 mole N_2, 1 mole O_2 and 3 moles CO_2 at a total pressure of 7 atm. The total number of moles of gas is 5. By substituting in the equations above:

$$x_{N_2} = \tfrac{1}{5} \qquad\qquad x_{O_2} = \tfrac{1}{5} \qquad\qquad x_{CO_2} = \tfrac{3}{5}$$

and

$$p_{N_2} = \tfrac{1}{5} \times 7\,\text{atm} = 1.40 \text{ atm,}$$

$$p_{O_2} = \tfrac{1}{5} \times 7\,\text{atm} = 1.40 \text{ atm,}$$

$$p_{CO_2} = \tfrac{3}{5} \times 7\,\text{atm} = 4.20 \text{ atm.}$$

In calculations such as this you should always check that the sum of the partial pressures is the same as the given total pressure.

The equilibrium constant in terms of partial pressures

The composition of a gaseous reaction mixture which has reached a state of dynamic equilibrium at a given temperature is determined by the equilibrium constant. Each of the gases in the mixture will exert a partial pressure as shown above. The equilibrium composition and hence the equilibrium constant can be defined in terms of these partial pressures rather than the molar concentrations used for K_c. Consider for example the reaction between nitrogen gas and hydrogen gas, used in the Haber process to produce ammonia:

$$N_2(g) + 3H_2(g) \rightleftharpoons 2NH_3(g)$$

If the total equilibrium pressure is P_t, and the partial pressures of the individual gases in the equilibrium mixture are p_{N_2}, p_{H_2} and p_{NH_3} respectively, then the equilibrium constant in terms of partial pressures, K_p, is given by the expression:

$$K_p = \frac{(p_{NH_3})^2}{(p_{H_2})^3 (p_{N_2})} \qquad \text{The units are pressure}^{-2}$$

The partial pressures are treated in exactly the same way as concentrations, so they are raised to a power according to the number of moles of gas in the equation for the reaction. The value of K_p describes the position of equilibrium and the concentrations of gases that can co-exist in an equilibrium. It does not generally have the same numerical value as K_c.

Gaseous dissociation

Thermal dissociation is the reversible breakdown of a substance into simpler substances on heating. It results in an equilibrium mixture in a closed system. One such reaction is the dissociation of dinitrogen tetroxide into nitrogen dioxide:

$$N_2O_4(g) \rightleftharpoons 2NO_2(g)$$

and this will be used to illustrate the foregoing principles.

The extent of such a dissociation can be given in terms of the percentage dissociation. Thus if we started with 100 mol of dinitrogen tetroxide which is 20% dissociated at equilibrium at a given temperature, there would be 80 mol left at equilibrium. The other 20 mol would have been converted to 40 mol of nitrogen dioxide. The equation shows that twice as many moles of NO_2 are formed from a given number of moles of N_2O_4. Putting the numbers of moles under the appropriate species in the equation, we have:

$$N_2O_4(g) \rightleftharpoons 2NO_2(g)$$

Numbers of moles initially: 100   0

Numbers of moles at equilibrium: 100 − 20   40

Total number of moles of gas at equilibrium   =   100 − 20 + 40 = 120

Therefore: $x_{N_2O_4} = \dfrac{80}{120} = 0.667$ and $x_{NO_2} = \dfrac{40}{120} = 0.333$

If the total equilibrium pressure is P_t, then the partial pressures are:

$$p_{N_2O_4} = 0.667 \times P_t \quad \text{and} \quad p_{NO_2} = 0.333 \times P_t$$

Now the value of K_p at this temperature can be calculated:

$$K_p = \frac{(0.333 \times P_t)^2}{0.667 \times P_t} = 0.166\, P_t$$

Instead of the percentage dissociation, the **degree of dissociation**, α, of a gas is frequently used. This is the fraction of the gas originally present that has dissociated. It is the percentage dissociation divided by 100 and is a number between 0 and 1. The degree of dissociation corresponding to 20% dissociation used in the example above is therefore 0.2.

The advantage in using the degree of dissociation to find the position of equilibrium is that the initial number of moles of N_2O_4 need not be known. It is as if the initial number of moles is 1.00. Consider again the dinitrogen tetroxide/nitrogen dioxide equilibrium:

$$N_2O_4(g) \rightleftharpoons 2NO_2(g)$$

Initially: 1   0

At equilibrium: $1 - \alpha$ 2α

Total: $1 - \alpha + 2\alpha = 1 + \alpha$

Therefore: $x_{N_2O_4} = \dfrac{1 - \alpha}{(1 + \alpha)}$ $x_{NO_2} = \dfrac{2\alpha}{(1 + \alpha)}$

If the gas is 20% dissociated at the given temperature, $\alpha = 0.2$.

Therefore: $x_{N_2O_4} = \dfrac{0.8}{1.2} = 0.667$ and $x_{NO_2} = \dfrac{0.4}{1.2} = 0.333$

The remainder of the calculation is as before.

Heterogeneous equilibria

If calcium carbonate is heated to a sufficient temperature, it decomposes to calcium oxide and carbon dioxide. In an open vessel, such as a test-tube, a blast furnace, or a lime-kiln, the carbon dioxide is lost and the reaction goes to completion. However, in a closed vessel the reaction is an equilibrium:

$$CaCO_3(s) \rightleftharpoons CaO(s) + CO_2(g)$$

Using the principles already covered, we can write an equilibrium constant for this reaction.

What is meant, though, by the concentration of a solid? The concentration of a pure solid is the number of moles divided by its volume (you may remember that mass in g = concentration, g cm^{-3} × volume, cm^3). It is no different from any other concentration – except that, since a pure solid has a constant density, it also has a constant concentration. Provided there is some discrete calcium carbonate and some calcium oxide present in the system, i.e. it does not form a solid solution, the concentration of each is constant. The equation for K_c does not need the concentrations of any solids to be included, so

$$K_c = [CO_2]$$

Factors affecting the position of equilibrium

There are three factors which can be changed that may alter the position of equilibrium:

- concentration
- pressure
- temperature

The way in which the position of equilibrium alters, if at all, can be deduced from the expression for K_c or K_p, or in the case of temperature changes, from the van't Hoff equation.

The effect of change in concentration at constant temperature

Consider a system in equilibrium and represented by the equation:

$$n\mathrm{A} + m\mathrm{B} \rightleftharpoons p\mathrm{P} + y\mathrm{Q}$$

$$K_c = \frac{[\mathrm{P}]^p[\mathrm{Q}]^y}{[\mathrm{A}]^n[\mathrm{B}]^m}$$

If the concentration of A is increased by adding more A to the system, then the value of the denominator will increase and the quotient is no longer K_c. Hence a new equilibrium position will be established, by reaction of A and B, in which the concentrations of P and Q will be greater than in the original equilibrium, so that once more the quotient is K_c. The position of equilibrium has been moved to the right in order to reduce the concentration of A.

The same effect would be produced if the concentration of B was to be increased or if the concentrations of either P or Q were reduced by removal from the system.

Similarly, increasing the concentration of P or Q, or decreasing the concentration of A or B would establish a new equilibrium position further to the left. In all cases, the new equilibrium concentrations are such as to maintain the value of K_c constant.

The effect of change in pressure at constant temperature

Pressure change can only affect a gaseous equilibrium, and then only if there is a change in the total numbers of moles of gas on going from one side of the equilibrium to the other.

For example, for the reaction:

$$2\mathrm{SO}_2(\mathrm{g}) + \mathrm{O}_2(\mathrm{g}) \rightleftharpoons 2\mathrm{SO}_3(\mathrm{g})$$

$$K_p = \frac{p(\mathrm{SO}_3)^2}{p(\mathrm{SO}_2)^2\, p(\mathrm{O}_2)}$$

$$= \frac{x(\mathrm{SO}_3)^2}{x(\mathrm{SO}_2)^2\, x(\mathrm{O}_2)\, P_t}$$

Increasing pressure increases the denominator. Thus $x(SO)_3$ must increase to restore the quotient to the value of K_p. A new equilibrium position would be established further to the right: that is, more SO_3 would be formed and less SO_2 and O_2 would be present. Note that the new pressure is the *equilibrium* total pressure.

In a reaction such as:

$$N_2(g) + O_2(g) \rightleftharpoons 2NO(g)$$

pressure changes would not affect the equilibrium position, since there is no change in the number of moles of gas as the reaction proceeds: there is no total pressure term in the expression for K_p.

The effect of change in temperature

In order to predict the effect of temperature changes on the equilibrium position, it is necessary to know whether the enthalpy change for the reaction is positive or negative. Since the reactions are reversible, you must remember that if H is positive for the reaction from left to right, then it will have the same value with the sign changed for the reaction from right to left.

The equation describing the behaviour is due to J.H. van't Hoff:

$$\ln \frac{K_2}{K_1} = \frac{\Delta H}{R} \left(\frac{1}{T_1} - \frac{1}{T_2} \right)$$

where K_1 and K_2 are the equilibrium constants at temperatures T_1 and T_2 and H is the enthalpy change for the reaction.

Thus for any system in equilibrium, for example:

$$2SO_2(g) + O_2(g) \rightleftharpoons 2SO_3(g) \qquad H = -188 \text{ kJ mol}^{-1}$$

the negative sign for H is taken to apply to the *complete* reaction left to right unless stated otherwise.

Consider an increase in *equilibrium* temperature. This would lead to a new position of equilibrium which lies further to the left; that is, a new equilibrium mixture would be obtained which contains higher concentrations of SO_2 and O_2. This is seen easily from the van't Hoff equation:

$$\ln \frac{K_2}{K_1} = \frac{\Delta H}{R} \left(\frac{1}{T_1} - \frac{1}{T_2} \right)$$
$$(-) \qquad (-) \qquad (+)$$

$\frac{\Delta H}{R}$ is negative because the reaction is exothermic and the third term positive. The first term thus being negative means $K_2 < K_1$; so at higher temperature, the equilibrium lies more to the left. Decreasing the temperature would conversely result in a new equilibrium mixture containing more SO_3. This can be summarised as follows:
  • exothermic reactions, low temperatures favour the right-hand side;
  • endothermic reactions, high temperatures favour the right-hand side

Temperature change affects the position of equilibrium **because** the value of the equilibrium constant changes with a change in temperature.

The rate of attainment of equilibrium

The factors which affect the *position* of equilibrium also affect the *rate* at which the equilibrium is reached. Thus an increase in concentration, pressure (if gases are involved) and temperature will all increase the rate at which the equilibrium is established, and will affect the position of equilibrium.

A catalyst will increase the rate at which an equilibrium is established, but **will not affect the position of equilibrium**. This is because in providing a new route for the reaction of lower activation energy, both the forward and the reverse reactions are affected equally. Hence the rates of both forward and reverse reactions increase, but the position of equilibrium remains unaltered. Reaction profiles for catalysed and uncatalysed reactions are shown in Figure 3.2.

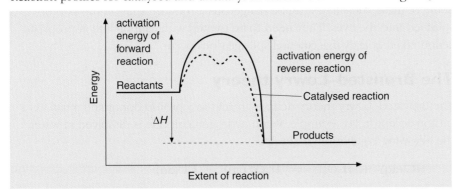

Fig. 3.2 Energy profile for a reversible reaction which is exothermic in the forward direction, showing the effect of a catalyst.

Questions

1 Calculate the mole fraction of each component in the following mixtures:
   (a) 1.6 mol of nitrogen and 0.4 mol of oxygen;
   (b) 0.5 mol of nitrogen, 0.7 mol of hydrogen, and 0.8 mol of ammonia.

2 Calculate the partial pressures of each of the gases in the preceding question if the total pressures are:
   (a) 2 atm;   (b) 500 kN m^{-2}.

3 At a given temperature a mixture of nitrogen and hydrogen, initially mixed in a 1:3 molar ratio, gives an equilibrium mixture containing 12% of ammonia at 5 atm total pressure.
   (a) Find the equilibrium constant for the reaction

   $$N_2(g) \; + \; 3H_2(g) \; \rightleftharpoons \; 2NH_3(g)$$

   at that temperature.
   (b) Raising the temperature of the equilibrium mixture by 200 K gives a new equilibrium constant of 1.00×10^{-6} atm^{-2} at the same total pressure. State with reasons whether the forward reaction is exo- or endothermic.

4 At a certain temperature and at 10 atm total pressure a sample of dinitrogen tetroxide is 30% dissociated. Find the equilibrium constant for the reaction

   $$N_2O_4(g) \; \rightleftharpoons \; 2NO_2(g)$$

   under these conditions.

5 At 1000 K, the equilibrium constant for the reaction

   $$H_2O(g) \; + \; C(s) \; \rightleftharpoons \; H_2(g) \; + \; CO(g)$$

   is 3.72 atm. Find the equilibrium partial pressures of each of the gases in the equilibrium mixture if the total pressure is 25 atm.

4

Acid–base equilibria

Fig. 4.1 Oranges, lemons and other citrus fruits contain citric acid.

Acids and bases are among the most familiar chemicals in a laboratory and many common household substances contain them. Probably the most common acid is vinegar, which is a very dilute solution of ethanoic acid (often still called acetic acid, CH_3COOH). Citrus fruits such as oranges, lemons and limes contain a very weak acid called citric acid, while car batteries contain sulphuric acid. Bases such as ammonia are present in many heavy duty cleaners and sodium hydroxide is found in many paint strippers and oven cleaners.

Some of these substances can be dangerous if they are spilt on the skin or splashed into the eyes. They need to be treated with caution and appropriate action taken quickly if accidental spillage does occur.

The Brønsted–Lowry theory

The Brønsted–Lowry theory defines an acid as a proton donor and a base as a proton acceptor. For example, when hydrogen chloride is dissolved in water, the following equilibrium is set up:

$$HCl(aq) + H_2O(l) \rightleftharpoons H_3O^+(aq) + Cl^-(aq)$$

In the forward reaction, the HCl is acting as an acid because it donates a proton (an H^+ ion) to the water which is acting as a base since it accepts a proton to become H_3O^+. In the reverse reaction the H_3O^+ ion acts as an acid by donating a proton to Cl^-, the latter acting as a base. Thus the equilibrium mixture consists of two acids and two bases and this must always be the case.

Conjugate acid–base pairs

The equation above can be split into two half-equations which show the proton transfer:

$$HCl - H^+ \rightleftharpoons Cl^-$$
Acid 1   Base 1

$$H_2O + H^+ \rightleftharpoons H_3O^+$$
Base 2   Acid 2

This clearly shows that when a species loses a proton, the product has to be a base since the process is reversible. This linkage of an acid to a base by the transfer of a single proton is recognised by use of the word '**conjugate**'. Thus Cl^- is the **conjugate base** of HCl and HCl the **conjugate acid** of Cl^-. Similarly, H_3O^+ and H_2O are a conjugate acid–base pair. Labelling equations as shown above (and below) is usually taken to be an adequate indication of the conjugate acid–base link.

Fig. 4.2 Typical household products containing alkalis, which are water-soluble bases.

Acids that have a single proton to donate are called **monoprotic** while those with two or three protons to donate are called **diprotic** and **tripotic**, respectively. (Alternatively, they are called monobasic, dibasic and tribasic acids.) Care must be taken with the use of the word 'conjugate' in these cases. For example, for the diprotic acid sulphuric acid, ionisation in water occurs in two stages:

$$H_2SO_4 \; + \; H_2O \; \rightleftharpoons \; H_3O^+ \; + \; HSO_4^-$$
$$\text{Acid 1} \quad \text{Base 2} \quad \text{Acid 2} \quad \text{Base 1}$$

$$HSO_4^- \; + \; H_2O \; \rightleftharpoons \; H_3O^+ \; + \; SO_4^{2-}$$
$$\text{Acid 3} \quad \text{Base 2} \quad \text{Acid 2} \quad \text{Base 3}$$

The behaviour of the hydrogen sulphate ion is different in the two equations, illustrating clearly that 'conjugate' is a relative term and links a *specific pair* of acids and bases. Thus HSO_4^- is the conjugate base of H_2SO_4 but it is the conjugate acid of SO_4^{2-}.

If appropriate molar quantities of sodium hydroxide are added to the sulphuric acid/water mixture above, two different salts can be formed, Na_2SO_4 (sodium sulphate) and $NaHSO_4$ (sodium hydrogen sulphate). This latter salt is quite acidic in aqueous solution, as indicated by the second equilibrium above which lies well over to the right. Hence it is used in many powder types of toilet cleaner, its function being to remove limescale ($CaCO_3$) from the lavatory bowl.

Fig. 4.3 Toilet cleaners often contain sodium hydrogen sulphate.

Acidic solutions

It is important to recognise the difference between the term **acid** and what is meant by an **acidic solution**. All too frequently the terms are taken to be synonymous, an acidic solution often being loosely referred to as an acid. For example, the substance HCl is a covalently bonded molecule and acts as an acid as it dissolves in water and donates its proton to the water molecule.

The resulting solution is acidic because it contains the hydrated hydrogen ion which is usually represented as H_3O^+ (or, more precisely, the solution is acidic because the concentration of H_3O^+ ions is greater than the concentration of OH^- ions as will be shown later in this chapter). Such acidic solutions therefore depend on the presence of water as the solvent and they have certain properties in common, such as:

- they react with bases to form salts
- they react with carbonates (which are of course bases) to produce carbon dioxide gas (the best test for an acidic solution)
- they react with many metals to give hydrogen
- they will produce a certain colour with acid–base indicators and have a pH of less than 7 at 25°C (see below).

A solution of HCl in a different solvent that cannot accept protons, such as methylbenzene, does not show any of these acidic properties.

Bases

The same principles can be applied to the behaviour of bases. Ammonia, for example, is a covalent molecule, but when dissolved in water the following equilibrium is set up:

$$NH_3(aq) + H_2O(l) \rightleftharpoons NH_4^+(aq) + OH^-(aq)$$
$$\text{Base 1} \qquad \text{Acid 2} \qquad \text{Acid 1} \qquad \text{Base 2}$$

The ammonia is acting as a base by accepting a proton from water. Water is acting as an acid, in contrast to its behaviour in the previous sections.

Again, the distinction needs to be drawn between a base and an alkaline solution. An alkaline solution contains a higher concentration of $OH^-(aq)$ ions than $H_3O^+(aq)$ ions. Thus the solution formed when ammonia dissolves in water is alkaline because of the OH^- ions produced, and NH_3 is acting as a base by accepting a proton.

In all acid–base equilibria the transfer of a proton can occur only if the proton can form a dative or coordinate bond with the species accepting it. Hence the species accepting it must have a lone pair of electrons which can be donated to form this dative bond.

There are other definitions of acids and bases, but they need not concern us here.

The effect of the solvent on acid–base behaviour

In solvents other than water, the behaviour of substances which are normally regarded as acids can be considerably modified. For example, ethanoic acid reacts with water, giving an acidic solution:

$$CH_3COOH(l) + H_2O(l) \rightleftharpoons CH_3COO^-(aq) + H_3O^+(aq)$$
$$\text{Acid 1} \qquad \text{Base 2} \qquad \text{Base 1} \qquad \text{Acid 2}$$

If, however, the ethanoic acid were to be dissolved in concentrated HCl the following equilibrium would be set up, in which the ethanoic acid behaves as a base.

$$CH_3COOH + HCl \rightleftharpoons CH_3COOH_2^+ + Cl^-$$
$$\text{Base 1} \qquad \text{Acid 2} \qquad \text{Acid 1} \qquad \text{Base 2}$$

Similarly, a mixture of concentrated nitric acid and concentrated sulphuric acid leads to some very surprising behaviour, with the concentrated nitric acid acting as a base:

$$HNO_3 + H_2SO_4 \rightleftharpoons H_2NO_3^+ + HSO_4^-$$
$$\text{Base 1} \qquad \text{Acid 2} \qquad \text{Acid 1} \qquad \text{Base 2}$$

This emphasises the point that the terms 'acid' and 'base' are *relative*, not absolute.

Measuring acidity: the pH scale

The acidity of a solution depends on the concentration of H_3O^+ ions and is measured on the pH scale, pH being defined as :

$$pH = -\log_{10}([H_3O^+(aq)]/\text{mol dm}^{-3})$$

The concentration is divided by the concentration unit to give a pure number; logarithms cannot be taken of physical quantities.

The use of this scale does do away with awkward numbers, particularly negative powers of ten.

Note:
- the negative sign in the definition means that the pH *decreases* as the hydrogen ion concentration *increases*
- a change of one unit on the pH scale corresponds to a tenfold change in the hydrogen ion concentration (since a $\log_{10}$ scale is used).

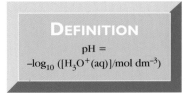

Thus if the hydrogen ion concentration in a solution is decreased ten times, the pH will increase by one unit.

The mathematical relationship between pH and the hydrogen ion concentration is such that if the pH is a whole number, then the conversion to the corresponding hydrogen ion concentration is very simple. If the pH is x, then the hydrogen ion concentration is 10^{-x} mol dm^{-3}. The relationship between the two quantities is shown in the following scale. Note that pH rises as $[H_3O^+]$ falls.

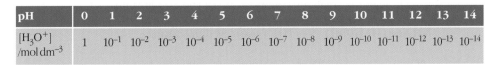

pH	0	1	2	3	4	5	6	7	8	9	10	11	12	13	14
$[H_3O^+]$ /mol dm^{-3}	1	10^{-1}	10^{-2}	10^{-3}	10^{-4}	10^{-5}	10^{-6}	10^{-7}	10^{-8}	10^{-9}	10^{-10}	10^{-11}	10^{-12}	10^{-13}	10^{-14}

This is the range of pH which is most commonly used. However, pH values of less than 0 and values of greater than 14 are possible, being limited only by the solubility of the strongest acid and the strongest base.

The hydrogen ion concentration in pure water is 1×10^{-7} mol dm^{-3} at 25°C, hence pure water has a pH of 7 at this temperature. Since pure water is the end product of the neutralisation of a strong acid and a strong base, a pH of 7 is regarded as defining a neutral solution **at 25°C**. Solutions with a pH of less than 7 are therefore acidic and those with a pH greater than 7 are alkaline, but only at 25°C.

Neutral solutions in general (i.e. at any temperature) are those where $[H_3O^+] = [OH^-]$. For water at 0°C, this corresponds to pH 7.5, and at 100°C, 6.1. The dissociation of water is an endothermic process, so this relationship of pH to temperature follows from standard equilibrium considerations covered earlier (page 34).

Strengths of acids

Strong and weak acids

The 'strength' of an acid is a term used to indicate the amount of ionisation which occurs when the acid is dissolved in water. In the case of hydrogen chloride, HCl is a covalent molecule which forms ions when dissolved in water:

$$HCl(aq) + H_2O(l) \rightleftharpoons H_3O^+(aq) + Cl^-(aq)$$

The equilibrium lies so far to the right that the acid can be considered to be completely ionised and it is said to be a strong acid. Other examples of strong acids are HNO_3 and (for its first ionisation only) H_2SO_4, both of which are assumed to be fully ionised in aqueous solution.

With an acid such as ethanoic acid, the equilibrium:

$$CH_3COOH(l) + H_2O(l) \rightleftharpoons CH_3COO^-(aq) + H_3O^+(aq)$$

lies much further to the left and the solution will contain CH_3COOH molecules as well as $CH_3COO^-(aq)$ and $H_3O^+(aq)$ ions. Ethanoic acid is therefore designated a 'weak' acid (being about 2% ionised). The terms 'strong' and 'weak' are obviously only comparative. It is difficult to know where to draw the line between them since acids can range from being 1 or 2% ionised to 100% ionised. A quantitative measure of acid strength is discussed in the next section.

Similarly, there are strong and weak bases. The hydroxides of the metals of Groups 1 and 2 of the Periodic Table are strong and are considered to be 100% ionised in aqueous solution. Ammonia, however, is a weak base because the ionisation in water is small, i.e. the position of the equilibrium:

$$NH_3(aq) + H_2O(l) \rightleftharpoons NH_4^+(aq) + OH^-(aq)$$

lies well over to the left. There are, however, sufficient hydroxide ions present to precipitate many metal hydroxides from solutions of the metal salt, a feature used in qualitative analysis.

The acid dissociation constant

The strength of an acid can be defined quantitatively by measuring the equilibrium constant, which determines the position of equilibrium. Consider, for example, a monoprotic acid HA which dissociates in water:

$$HA(aq) + H_2O(l) \rightleftharpoons H_3O^+(aq) + A^-(aq)$$

The equilibrium constant, K_c, for this is:

$$K_c = \frac{[H_3O^+][A^-]}{[HA][H_2O]}$$

Since the concentration of $H_2O(l)$ is effectively constant, the water being in large excess, $[H_2O]$ can be incorporated into the value for K_c so that it becomes a new constant which is given the symbol K_a:

$$K_a = \frac{[H_3O^+][A^-]}{[HA]}$$

> **DEFINITION**
>
> The **acid dissociation constant** for a monoprotic acid HA:
> $$K_a = \frac{[H_3O^+][A^-]}{[HA]}$$

K_a is the **acid dissociation constant**.

Note:
- the expression must not have $[H_2O]$ on the bottom line
- K_a is an equilibrium constant; it is temperature dependent
- K_a has units of $mol\,dm^{-3}$
- the value of K_a determines the position of equilibrium. The greater the value of K_a, the stronger the acid HA.

Thus a quantitative comparison of the strengths of acids *when dissolved in water* can be made. It is important to realise that these values of K_a are only valid for solutions in water where the base is the same (i.e. H_2O) no matter which acid is dissolved in it. The strengths will be different in other solvents.

Some values of K_a for some common acids are given in Table 4.1.

Table 4.1 K_a *values for some weak acids*

Acid	$K_a/mol\,dm^{-3}$
ethanoic acid	1.70×10^{-5}
chloroethanoic acid	1.38×10^{-3}
dichloroethanoic acid	5.13×10^{-2}
benzoic acid	6.30×10^{-5}
nitrous acid	5.00×10^{-4}

For polyprotic acids, a K_a value can be written for each stage of the ionisation, called K_1, K_2, etc. These show that the ionisation gets progressively weaker. The values for phosphoric(V) acid are shown in Table 4.2.

Table 4.2 K_a *values for the successive ionisations of phosphoric(V) acid*

	$K_a/mol\,dm^{-3}$
$H_3PO_4 + H_2O \rightleftharpoons H_3O^+ + H_2PO_4^-$	8.0×10^{-3}
$H_2PO_4^- + H_2O \rightleftharpoons H_3O^+ + HPO_4^{2-}$	6.3×10^{-8}
$HPO_4^{2-} + H_2O \rightleftharpoons H_3O^+ + PO_4^{3-}$	4.0×10^{-13}

Ionisation of water

Water, no matter how pure it is made, always ionises to a very small extent:

$$H_2O(l) + H_2O(l) \rightleftharpoons H_3O^+(aq) + OH^-(aq)$$

and this is itself an acid–base equilibrium. This ionisation is particularly interesting in that the water produces its own conjugate acid and its own conjugate base at the same time.

Application of the equilibrium law to this leads to the following expression for the equilibrium constant K_c:

$$K_c = \frac{[H_3O^+(aq)][OH^-(aq)]}{[H_2O]^2}$$

ACID–BASE EQUILIBRIA

The denominator $[H_2O]^2$ is effectively constant since water is in large excess and so the top line of the above expression must also be constant. Hence:

$$[H_3O^+(aq)][OH^-(aq)] = \text{constant} = K_w$$

K_w is called the **ionic product of water**.

Since K_w is an equilibrium constant, it is temperature dependent. K_w has units of $mol^2\,dm^{-6}$. Its numerical value is about $1 \times 10^{-14}\,mol^2\,dm^{-6}$ at 298 K, but the value varies with temperature. At 0°C it is $1 \times 10^{-15}\,mol^2\,dm^{-6}$, at 100°C $5.4 \times 10^{-13}\,mol^2\,dm^{-6}$.

In water, the concentration of the H_3O^+ ion must be the same as that of the OH^- ion. Thus:

$$[H_3O^+(aq)] = [OH^-(aq)] = \overline{1 \times 10^{-14}} = 1 \times 10^{-7}\,mol\,dm^{-3}$$

When acids or alkalis are dissolved in water there will be different concentrations of $H_3O^+(aq)$ and $OH^-(aq)$ but the product of these two concentrations is always equal to the value of K_w at that temperature.

If, for example, a solution is made which contains $0.1\,mol\,dm^{-3}$ HCl, this will provide a concentration of $H_3O^+(aq)$ of $0.1\,mol\,dm^{-3}$ (assuming that the acid ionises completely). Thus the water ionisation is suppressed, i.e. the equilibrium moves to the left, but the value of K_w must still be maintained. Assuming that the $[H_3O^+(aq)]$ from the water is now so small compared to that from the acid that it is negligible:

$$[H_3O^+(aq)][OH^-(aq)] = K_w = 10^{-14}\,mol^2\,dm^{-6}$$

$$\therefore\ 0.1\,mol\,dm^{-3} \times [OH^-] = 10^{-14}\,mol^2\,dm^{-6}$$

$$\therefore\ [OH^-] = 10^{-13}\,mol\,dm^{-3}$$

There is, therefore, a very small, but definite, concentration of hydroxide ions in a solution of an acid. Application of the same principles leads to the conclusion that solutions of alkalis have a small, but definite, hydrogen ion concentration.

Calculating the pH of strong acids

For purposes of calculation, it is assumed that all strong acids are completely ionised. Hence the hydrogen ion concentration is obtained directly from the molarity of the acid. For example, in a solution of hydrochloric acid having a concentration of $0.1\,mol\,dm^{-3}$, the acid ionises completely:

$$HCl(aq) + H_2O(l) \rightarrow H_3O^+(aq) + Cl^-(aq)$$

$$\text{so,}\quad [H_3O^+] = 0.1\,mol\,dm^{-3}, \text{ and } \log_{10}0.1 = -1;$$

$$\therefore\ -\log_{10}0.1 = 1 \qquad \text{and so} \qquad pH = 1$$

With diprotic acids such as sulphuric acid, H_2SO_4, it might be thought that the hydrogen ion concentration is double, so that a $0.1\,mol\,dm^{-3}$ solution of sulphuric acid would have $[H_3O^+] = 0.2\,mol\,dm^{-3}$. This is not so. The second ionisation of sulphuric acid

$$HSO_4^-(aq) + H_2O(aq) \rightleftharpoons SO_4^{2-}(aq) + H_3O^+(aq)$$

is fairly weak, with $K_a = 0.01$ mol dm^{-3}. It is not very difficult to show that this ionisation is suppressed by the first ionisation which gives a high $[H_3O^+]$. It contributes so little to the hydrogen ion concentration that the pH of 0.1 mol dm^{-3} sulphuric acid is about 0.98, not very different from HCl(aq) of the same concentration with pH = 1.

Calculating of the pH of strong bases

Strong bases are also assumed to be completely ionised. The hydroxide ion concentration is therefore easily obtained from the molarity of the base. For example a 0.3 mol dm^{-3} solution of sodium hydroxide is ionised:

$$NaOH(aq) \rightarrow Na^+(aq) + OH^-(aq)$$

So, $[OH^-] = 0.3$ mol dm^{-3}

The hydrogen ion concentration which is present in this solution is determined by the value of K_w for water. If this is taken to be 10^{-14} mol^2 dm^{-6} at 298 K, then the hydrogen ion concentration at this temperature is given by:

$$[H_3O^+][OH^-] = 10^{-14} \text{ mol}^2 \text{ dm}^{-6}$$

$$\therefore 0.3 \text{ mol dm}^{-3} \times [H_3O^+] = 10^{-14}$$

$$[H_3O^+] = \frac{1 \times 10^{-14} \text{ mol}^2 \text{ dm}^{-6}}{0.3 \text{ mol dm}^{-3}} = 3.33 \times 10^{-14} \text{ mol dm}^{-3}$$

$$\therefore pH = -\log_{10}(3.33 \times 10^{-14}) = 13.5$$

Calculating the pH of weak acids

More information is required to calculate the pH of weak acids. As well as the molarity of the solution, it is necessary to know the degree of ionisation of the acid, or its K_a value. For example, calculate the pH of a 0.1 mol dm^{-3} solution of ethanoic acid at 298 K given that $K_a = 1.7 \times 10^{-5}$ mol dm^{-3} at this temperature. The equilibrium:

$$CH_3COOH(aq) + H_2O(l) \rightleftharpoons CH_3COO^-(aq) + H_3O^+(aq)$$

is often represented in a simpler form for the purposes of calculation:

$$CH_3COOH(aq) \rightleftharpoons CH_3COO^-(aq) + H^+(aq)$$

$$K_a = \frac{[CH_3COO^-(aq)][H^+(aq)]}{[CH_3COOH(aq)]}$$

The ethanoate ions and hydrogen ions must be produced in equal concentration, so:

$$K_a = \frac{[H^+(aq)]^2}{[CH_3COOH(aq)]}$$

For a weak acid, the ionisation is assumed to be so small that it is negligible. Hence $[CH_3COOH(aq)]$ is assumed to be 0.1 mol dm^{-3}.

$$\therefore K_a = 1.7 \times 10^{-5} \text{ mol dm}^{-3} = \frac{[H^+(aq)]^2}{0.1 \text{ mol dm}^{-3}}$$

$$[H^+(aq)]^2 = 0.1 \times 1.7 \times 10^{-5}\,\text{mol}^2\,\text{dm}^{-3}$$
$$[H^+(aq)] = \overline{0.1 \times 1.7 \times 10^{-5}\,\text{mol}^2\,\text{dm}^{-6}} = 1.3 \times 10^{-3}\,\text{mol}\,\text{dm}^{-3}$$
$$\therefore pH = -\log_{10}(1.3 \times 10^{-3}) = 2.88$$

In the above calculations it is important that you are able to carry out the calculations in reverse; that is, given the pH value you should be able to calculate the hydrogen ion concentration for any strong acid or base or calculate the K_a value for a weak acid.

For example, if a solution of hydrochloric acid has a pH of 3, calculate the hydrogen ion concentration and the molarity of the acid.

$$pH = 3, \therefore [H_3O^+] = \text{antilog}\,(-3) = 1 \times 10^{-3}\,\text{mol}\,\text{dm}^{-3}$$

Assuming the acid to be 100% ionised, and since the acid is monoprotic, the molarity of the acid must also be $1 \times 10^{-3}\,\text{mol}\,\text{dm}^{-3}$.

It is important to appreciate the following points.

1 The difference between the strength of an acid and the concentration of a solution:

Strength refers to the extent of ionisation or dissociation of the acid into hydrogen ions, while *concentration* refers to the number of moles of acid dissolved in a given volume of water. It is therefore quite possible to have a concentrated solution of a weak acid or a dilute solution of a strong acid.

2 The pH value of a solution by itself is not a measure of the strength of an acid:

The pH of a solution is simply related to the hydrogen ion concentration in the solution. Some knowledge of the molar concentration of the acid solution is needed before a judgement can be made about the strength of the acid from its pH value. It was shown in an example above that the pH of a $0.1\,\text{mol}\,\text{dm}^{-3}$ solution of ethanoic acid is 2.88. A $0.001\,\text{mol}\,\text{dm}^{-3}$ solution of hydrochloric acid would have a pH of 3.0. This does not mean that ethanoic acid is a stronger acid than hydrochloric acid, but simply that, for the concentrations of solution given, it has a greater hydrogen ion concentration and hence a lower pH value.

Acid–base titrations

A *titration* is a quantitative technique in which the volume of one solution required to completely react with a known volume of another solution is accurately measured. This requires us to have some means of knowing when the reaction is complete, that is, when the acid and base are present in stoichiometric proportions as shown by the equation. This is called the **end point** or **equivalence point** for the titration.

Certain indicators, which have different colours depending on the pH of the solution in which they are placed, can be used. In order to understand how these indicators work, it is necessary first of all to understand the way in which the pH changes when an acid is gradually added to a base, or vice versa, until in excess.

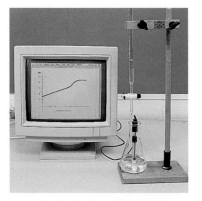

Fig. 4.4 Producing a pH curve from an acid–base titration.

pH changes during titrations

The change in pH during a titration can be followed by using a pH meter to measure the pH after each addition of acid or base. The graph of pH against volume of acid or base added is then plotted to give a so-called titration curve. The shape of the curve obtained is dependent to some extent on the strengths of the acid and base used, and these are shown in Figure 4.5 for various combinations of strong and weak acids and bases.

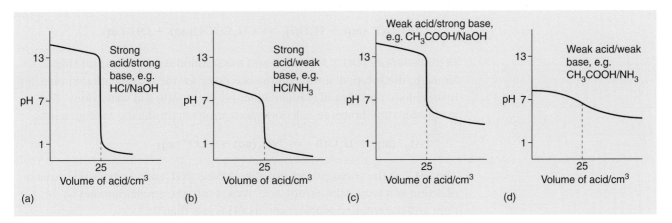

Fig. 4.5 Titration curves for various solutions of acids and bases. Each curve relates to the volume of 0.10 mol dm^{-3} aqueous acid added to 25 cm^3 of 0.10 mol dm^{-3} aqueous base.

Note that the amount of acid required to neutralise the base in all these cases does **not** depend on the strength of the acid but on the stoichiometric equation for the reaction.

Buffering effects during titrations (see also page 49)

As can be seen in Figure 4.5 there is a large change in pH for a very small addition of acid around the end point of the titration, in all cases except that for weak acid/weak base. The pH change is about 12 units for strong acid/strong base titrations but only about 6 units for weak acid/strong base and for strong acid/weak base. The change is not so great in these cases since the solution just before or after the end point consists of a buffer solution which resists such large changes in pH.

In the case of the titration of CH$_3$COOH solution with NaOH solution, the solution after the end point contains sodium ethanoate and ethanoic acid which is a buffer solution, as described above (see also the section on Buffer solutions later). Similarly, in the case of HCl reacting with NH$_3$, the solution just before the end point consists of the weak base ammonia and its salt ammonium chloride. Hence the solution is again buffered.

Weak acid/weak base titrations such as CH$_3$COOH solution with NH$_3$ solution contain a buffer solution both before the end point (NH$_3$ + ammonium ethanoate) as well as after the end point (CH$_3$COOH + ammonium ethanoate), hence the large change in pH does not occur, and it is not possible to perform such titrations using an acid–base indicator. Measurement of the electrical conductivity of the solution, or of heat changes, will enable a titration to be performed in such cases.

The pH at the equivalence point

It can be seen from Figure 4.5 that the pH at the equivalence point of a titration is not always 7. In the case of a weak acid /strong base titration such as CH_3COOH with NaOH, the pH at the equivalence point is greater than 7. The only substances present in the solution at the end point are the salt, sodium ethanoate, and water. The ethanoate anion is the conjugate base of a weak acid and hence a reasonably strong base, capable of reacting with water. This is **salt hydrolysis**:

$$CH_3COO^- (aq) + H_2O(l) \rightleftharpoons CH_3COOH(aq) + OH^-(aq)$$

In this solution, $[OH^-] > [H_3O^+]$ and the solution has a pH greater than 7. Similarly, the solution at the equivalence point for the titration of ammonia and hydrochloric acid should contain ammonium chloride and water only. The ammonium ion, however, will undergo hydrolysis, producing hydrogen ions:

$$NH_4^+(aq) + H_2O(l) \rightleftharpoons NH_3(aq) + H_3O^+(aq)$$

The NH_4^+ is the conjugate acid of a weak base (NH_3) and is therefore capable of acting as a reasonably strong acid. As a result, the equilibrium lies to the right and the solution is acidic and its pH is less than 7.

Determination of K_a for a weak acid

If a weak acid is titrated with a strong base, the titration curve is as shown in Figure 4.6. Suppose that this is for the titration of ethanoic acid with sodium hydroxide:

$$CH_3COOH + NaOH \rightarrow CH_3COONa + H_2O$$

(Note that in this case the base is being added to the acid.)

When the neutralisation is half complete, i.e. half the required amount of base to reach the equivalence point has been added, the concentration of the salt sodium ethanoate is the same as that of the ethanoic acid. Thus:

$$K_a = \frac{[H^+][CH_3COO^-]}{[CH_3COOH]}$$

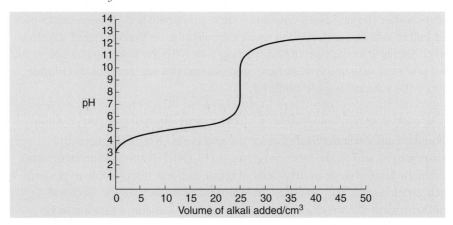

Fig. 4.6 Titration curve for sodium hydroxide added to ethanoic acid.

and if $[CH_3COO^-] = [CH_3COOH]$ then $K_a = [H^+]$ and pH = pK_a. Therefore the value of the pH at the half-neutralisation point is equal to pK_a.

Acid–base indicators

Behaviour of acid–base indicators

Acid–base indicators are complex organic molecules, but their behaviour is relatively simple to understand. They are all very weak acids in which the conjugate base is a different colour from the acid itself (Figure 4.7). Their dissociation can be represented by the equilibrium:

$$HInd + H_2O \rightleftharpoons H_3O^+ + Ind^-$$

Colour 1   Colour 2

The strength of an indicator as an acid will depend on the K_a value for the indicator. This is called K_{Ind}:

$$K_{Ind} = \frac{[Ind^-][H_3O^+]}{[HInd]}$$

Note that $[H_2O]$ does not appear since K_{Ind} is just a particular version of K_a.

Addition of acid to this solution will push the equilibrium to the left and colour 1 will be seen. Addition of alkali will result in the equilibrium moving to the right since the OH^- ions will remove hydrogen ions from the equilibrium. Colour 2 will be seen. The indicator will show an intermediate colour when colour 1 and colour 2 are present in equal concentration, i.e. when $[Ind^-] = [HInd]$. In this situation:

$$K_{Ind} = \frac{[Ind^-][H_3O^+]}{[HInd]} \quad \text{and therefore} \quad K_{Ind} = [H_3O^+]$$

The indicator will show its intermediate colour at a pH value which is determined by the K_{Ind} value for the indicator. Since indicators have different K_{Ind} values, they will change colour at different pH values. The complete colour change from colour 1 to colour 2, or vice versa, requires a change of about 1.5 to 2 units of pH. This range is the **working range** of the indicator. The pH of the intermediate colour is at the midpoint of this range. Some values for common indicators are shown in Table 4.3.

Table 4.3 *The working range and colour changes of some indicators*

Indicator	pH range	Acid colour	Alkaline colour
methyl orange	3.2–4.5	red	yellow
phenolphthalein	8.2–10.0	colourless	magenta
bromothymol blue	6.0–7.0	yellow	blue

Choice of indicator for titrations

The indicator used for any acid–alkali titration should ideally change colour at the pH corresponding to the midpoint of the straight and nearly vertical portion of the titration curve. However, there is no serious loss of accuracy if the indicator changes colour anywhere on the straight portion of the titration curve since the

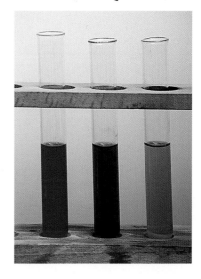

Fig. 4.7 The colouring material in red cabbage is an indicator. The red coloured extract shown in the left-hand tube is yellow with acid (right-hand tube), and blue/purple with alkali (middle tube).

pH change is so large for the addition of a very small amount of acid or base. For a strong acid/strong base titration, any of the indicators in Table 4.3 could be used. More care is needed, however, in the choice of indicator when either the acid or base is weak, since the pH range of the straight portion is smaller. Thus phenolphthalein could be used for a strong base/weak acid titration, whereas methyl orange could not. Similarly methyl orange would be much more suitable than phenolphthalein for strong acid/weak base titrations.

Titration curves for diprotic acids

A diprotic acid such as ethanedioic acid $H_2C_2O_4$ has two protons which can be successively replaced in what are effectively separate acid–base reactions.

$$H_2C_2O_4(aq) + NaOH(aq) \rightarrow NaHC_2O_4(aq) + H_2O(l) \qquad \text{Step 1}$$

$$NaHC_2O_4(aq) + NaOH(aq) \rightarrow Na_2C_2O_4(aq) + H_2O(l) \qquad \text{Step 2}$$

The titration curve for such a reaction shows two vertical portions, each corresponding to the completion of one of these steps, as shown in Figure 4.8.

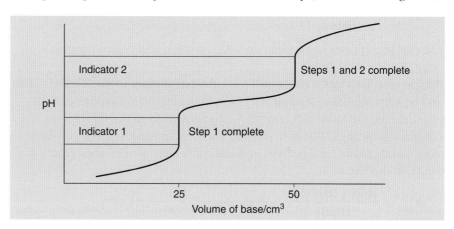

Fig. 4.8 Titration curve for the titration of 25 cm³ of 0.10 mol dm⁻³ ethanedioic acid with 0.10 mol dm⁻³ sodium hydroxide. Note the 'volume of base' on the x-axis; volume of acid was used in Fig.4.5.

An appropriate choice of indicator would allow either of the two end points to be identified. It is obviously important to know which end point has been identified when using the results of such titrations for calculation, since the equation used in the calculation must be the correct one for the indicator used.

Thus if indicator 1 is used, the equation for step 1 only is appropriate. If indicator 2 is used then the equations for step 1 and step 2 must be added together.

Acid strength and enthalpies of neutralisation

The enthalpy of neutralisation is defined as the enthalpy change when one mole of water is formed from reaction of an acid with a base. For example, it would be H for any of the following reactions:

$$HCl(aq) + NaOH(aq) \rightarrow NaCl(aq) + H_2O(l)$$

$$\tfrac{1}{2}H_2SO_4(aq) + NaOH(aq) \rightarrow \tfrac{1}{2}Na_2SO_4(aq) + H_2O(l)$$

$$CH_3COOH(aq) + NaOH(aq) \rightarrow CH_3COONa(aq) + H_2O(l)$$

The value of the enthalpy of neutralisation is remarkably constant at about $-57 \, kJ \, mol^{-1}$ when the acid and the base are both strong. This is not really surprising when you realise that if the acid and base are fully ionised, the ionic equation for all these reactions is the same, that is:

$$H^+(aq) + OH^-(aq) \rightarrow H_2O(l)$$

Hence we are simply measuring the enthalpy change for the same reaction. If either the acid or base is weak, however, some of the energy is absorbed in order to ionise the weak acid or base. Hence the amount of heat liberated is usually less than $-57 \, kJ \, mol^{-1}$.

For example, the enthalpy of neutralisation for ethanoic acid and sodium hydroxide is $-55.2 \, kJ \, mol^{-1}$. Assuming the base to be fully ionised, the following diagram illustrates the enthalpy changes occurring:

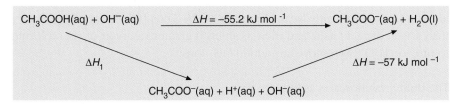

Application of Hess's law enables us to calculate the amount of energy (H_1) which has been required to ionise the weak acid :

$$-55.2 = H_1 + (-57)$$
$$\therefore H_1 = 57 - 55.2 = +1.8 \, kJ \, mol^{-1}$$

This is not a true figure for the full amount of energy required to ionise the acid since there will also be heat liberated on hydration of the ions once they have been formed. The value of $+1.8 \, kJ \, mol^{-1}$ is the sum of both these processes. Energy released on hydration of the ions is sometimes greater than the energy required to complete the ionisation of the acid or base, and this explains why the values of enthalpy of neutralisation of some acids and bases are more exothermic than $-57 \, kJ \, mol^{-1}$. Some values for other acids and bases are shown in Table 4.4.

Table 4.4 *The enthalpies of neutralisation of some acids and bases*

Reaction	$-H/kJ \, mol^{-1}$
HCl/NaOH	57.1
HNO_3/KOH	57.3
CH_3COOH/NaOH	55.2
HCl/NH_3	52.2
$HNO_3/^1/_2Ba(OH)_2$	58.2

Buffer solutions

The nature and function of buffer solutions

Buffer solutions are solutions of known pH which have the ability to resist changes in pH when contaminated by small amounts of acid or alkali. Buffer solutions are very important in certain situations where the pH of a solution must be maintained. Blood, for example, is buffered at pH 7.4 and a variation of only 0.5 in this pH could prove fatal. It is obviously important that any injections into the bloodstream, given for medical reasons, should also be buffered.

> **DEFINITION**
>
> **Buffer solutions** are solutions of known pH which have the ability to resist changes in pH when contaminated with small amounts of acid or alkali.

The simplest form of buffer solution is made using a solution containing a weak acid together with a salt of the same acid, for example, ethanoic acid, CH_3COOH, plus sodium ethanoate, CH_3COONa. The salt obviously has an important function which can be understood if we first consider what would happen in a solution of the acid on its own.

A weak acid HA will dissociate in solution as follows:

$$HA(aq) + H_2O(l) \rightleftharpoons H_3O^+(aq) + A^-(aq)$$

where $[H_3O^+(aq)] = [A^-(aq)]$ and both concentrations are small. The solution of HA could cope quite well with the addition of OH^- ions since they would combine with the H_3O^+ to form water. The equilibrium would move to the right to replace the H_3O^+ removed and hence the pH would be maintained nearly constant. Addition of more H_3O^+ ions, however, would push the equilibrium to the left by combining with A^- ions. The small concentration of A^- ions would very rapidly fall to zero; any further addition of H_3O^+ would not be removed and hence the pH would change rapidly.

The presence of the salt of the weak acid Na^+A^- gives a large reservoir of A^- ions and hence the reaction with added H_3O^+ ion can continue for much longer and a change in pH is thus resisted. A vital function of the salt is that the added A^- ions suppress the already weak ionisation of HA. Thus [HA] is actually greater than it is in the acid itself and there is an even greater reservoir of HA molecules to provide H_3O^+ to remove any added OH^- ions.

The buffer solution acts as follows:

• Added hydrogen ions are removed from solution by:

$$H_3O^+(aq) + A^-(aq) \rightleftharpoons HA(aq) + H_2O(l)$$

• Added hydroxide ions are removed from solution by the reaction:

$$HA(aq) + OH^-(aq)) \rightleftharpoons H_2O(l) + A^-(aq)$$

During these processes, neither [HA] nor [A⁻] changes by very much and so the hydrogen ion concentration does not change very much either: the pH remains almost constant.

Calculating the pH of a buffer solution

Consider a solution at 298 K which contains $0.1 \, mol \, dm^{-3}$ ethanoic acid and $0.2 \, mol \, dm^{-3}$ sodium ethanoate, the K_a value for ethanoic acid being $1.7 \times 10^{-5} \, mol \, dm^{-3}$ at this temperature. The calculation is made from the K_a expression:

$$K_a = \frac{[CH_3CO_2^-(aq)][H^+(aq)]}{[CH_3CO_2H(aq)]}$$

but in this situation, $[CH_3CO_2^-(aq)]$ and $[H^+(aq)]$ are *not* equal, since the salt has been added to the solution.

Two assumptions are made:
• The anion concentration from the acid is very small compared to that from the salt (which is assumed to be 100% ionised). The salt is considered to be the only source of the anion. So, in this case:

$$[CH_3COO^-(aq)] = [\text{sodium ethanoate}] = 0.2 \, mol \, dm^{-3}$$

- The presence of excess anion from the salt has suppressed the ionisation of the acid to such a low level that it can be considered negligible. So, in this case:

$$[CH_3COOH(aq)] = [\text{ethanoic acid}] = 0.1 \, \text{mol dm}^{-3}$$

$$1.7 \times 10^{-5} \, \text{mol dm}^{-3} \; = \; \frac{[H^+(aq)] \times 0.2 \, \text{mol dm}^{-3}}{0.1 \, \text{mol dm}^{-3}}$$

$$[H^+(aq)] \; = \frac{1.7 \times 10^{-5} \, \text{mol dm}^{-3} \times 0.1 \, \text{mol dm}^{-3}}{0.2 \, \text{mol dm}^{-3}} \; = \; 8.5 \times 10^{-6} \, \text{mol dm}^{-3}$$

$$pH = -\log_{10}(8.5 \times 10^{-6}) \; = \; 5.07$$

The effect on this buffer solution of adding hydrogen (or hydroxide) ions can also be calculated. Suppose, for example, that $1 \, \text{cm}^3$ of $1.0 \, \text{mol dm}^{-3}$ hydrochloric acid is added to $1 \, \text{dm}^3$ of the above buffer solution. The hydrogen ion concentration has been increased by $10^{-3} \, \text{mol dm}^{-3}$. This is removed by:

$$H_3O^+(aq) \; + \; A^-(aq) = HA(aq) + H_2O$$

The concentration of A^- will thus decrease by 0.001 to a new value of $0.2 - 0.001 = 0.199 \, \text{mol dm}^{-3}$. The concentration of HA will increase to $0.1 + 0.001 = 0.101 \, \text{mol dm}^{-3}$. The concentration of hydrogen ions that can co-exist with these new concentrations is calculated from:

$$[H^+(aq)] \; = \frac{1.7 \times 10^{-5} \, \text{mol dm}^{-3} \times 0.101 \, \text{mol dm}^{-3}}{0.199 \, \text{mol dm}^{-3}} \; = \; 8.63 \times 10^{-6} \, \text{mol dm}^{-3}$$

$$\therefore \; pH = -\log_{10}(8.63 \times 10^{-6}) = 5.06$$

The pH has changed by only 0.01 unit.

The addition of the same amount of acid to $1 \, \text{dm}^3$ of distilled water would reduce the pH from 7 to 3.

Organic Chemistry II

Further ideas in organic chemistry

The ideas that were developed in Unit 2 are expanded in this unit to cover a further range of important compounds and their inter-conversion. First we will revise some significant ideas on functional groups and isomerism.

Functional groups

Organic compounds in general have an atom or a group of atoms, the functional group, which determines the chemical reactions that are possible. Alkenes, for example, have a carbon–carbon double bond that defines the chemistry as largely (though not exclusively) that of addition reactions (Topic 2.2 of AS). In this Unit we will consider the functional groups listed in Table 5.1.

Table 5.1 *The compound types and functional groups considered in this Unit*

Compound type	Functional group	Compound type	Functional group
halogenoalkanes	R-**X** where X is Cl, Br or I	acid chlorides	R-**COCl**
Grignard reagents	R-**MgX** where X is a halogen atom	amines	R-**NH$_2$**
carboxylic acids	R-**COOH**	amides	R-**CONH$_2$**
aldehydes	R-**CHO**	nitriles	R-**CN**
ketones	R-**CO-R´**	amino acids	RCH(**NH$_3^+$**)**COO$^-$**

The functional group is in bold.

Isomerism

In Unit 2 the idea of isomerism was introduced, i.e. the common feature that a collection of atoms can often be joined together in perhaps several different ways. **Isomers** are therefore molecules that have the same molecular formula, but have different structural formulae.

Structural isomerism

In alkanes, the only way in which the structure can be changed is by rearranging the carbon chain, and the difference between straight-chain and branched-chain isomers will be familiar to you. Once another functional group has been introduced into the molecule, other possibilities arise. If, for example, we consider the amines having four carbon atoms, there are two that have a straight chain:

$$CH_3CH_2CH_2CH_2NH_2 \qquad CH_3CH_2CH(NH_2)CH_3$$

and two that have a branched chain:

$$\begin{array}{cc} CH_3 & CH_3 \\ | & | \\ CH_3CHCH_2NH_2 & CH_3C(NH_2)CH_3 \end{array}$$

The number of structural isomers that a compound can have may be large; it is more important that you recognise how to spot the possibility of isomerism rather than learn enormous numbers of different structures.

Geometric isomerism

Geometric isomerism arises because of restricted rotation about a carbon–carbon double bond. The restricted rotation is a consequence of the sideways overlap of the p orbitals that give the π bond; if the atoms at each end are rotated, the orbitals no longer overlap and the bond is broken. This process requires energy, and does not happen at room temperature. Heating can break the bond, though, and heating geometric isomers will sometimes cause their inter-conversion.

The requirement is not only for a C=C double bond, however; in the structure below it must be the case that a $\neq$ e and b $\neq$ d:

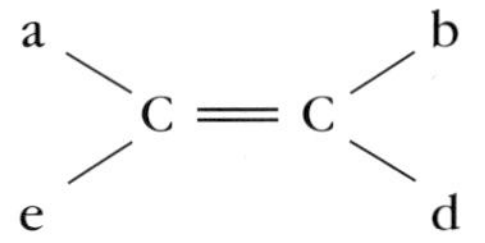

The inclusion of functional groups other than alkyl groups will make such isomerism more likely in compounds containing C=C bonds. An example is cinnamic acid, the *cis*-form of which is shown:

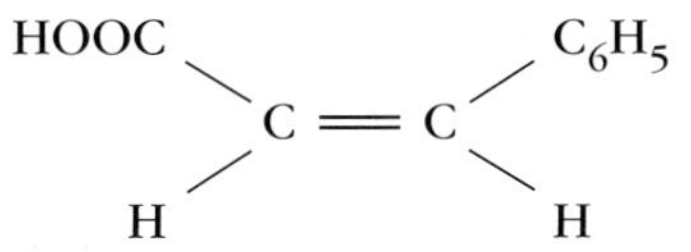

Some further examples are:

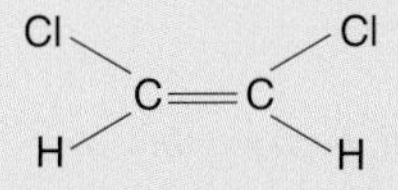

cis-1,2-dichloroethene

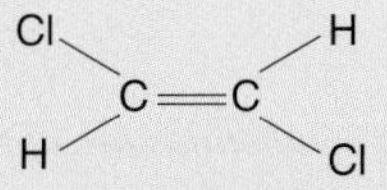

trans-1,2-dichloroethene

There is also a structural isomer, 1,1-dichloroethene.

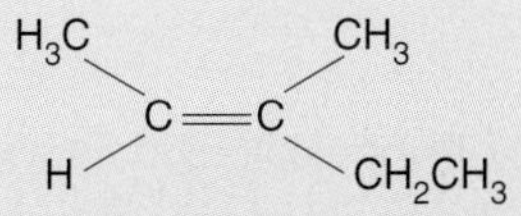

cis-3-methylpent-2-ene

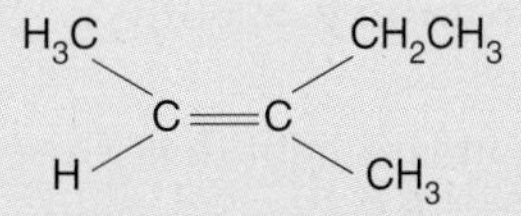

trans-3-methylpent-2-ene

Geometric isomerism is one aspect of **stereoisomerism**, the isomerism that arises because of different *orientations* of groups in space rather than the compounds having a fundamentally distinct carbon chain. Again there will be many possibilities once the number of functional groups in a molecule increases, but if the molecule's structure fulfils the criterion shown above, it will have geometric isomers.

> ### DEFINITION
>
> **Geometric isomerism** occurs when there is restricted rotation about the carbon–carbon double bond and the groups on a given carbon atom are not the same.

Optical isomerism

Optical isomerism is found where molecules have mirror-image isomers that are not superimposable on the original compound. Such molecules are **chiral**; the commonest origin of chirality is a carbon atom having four different groups attached to it, called a **chiral centre**. However, molecules with two such chiral centres may not be chiral as a whole; indeed they will not be if one chiral centre is the mirror image of the other. It is also possible to have chirality in molecules that do not have chiral centres, for example if the molecule is helical. The *sole* criterion (the necessary and sufficient condition) for chirality is the existence of non-superimposable mirror images. The examples shown are for 2-methylbutanal, an aldehyde; and for alanine, a naturally occurring amino acid found in protein.

> ### DEFINITION
>
> A **chiral molecule** is one which has an isomer that is a mirror image of itself, the two being non-superimposable.

Optical activity

Chiral molecules possess the property that, if plane-polarised monochromatic light is shone through them, the plane of polarisation is **rotated**. Molecules

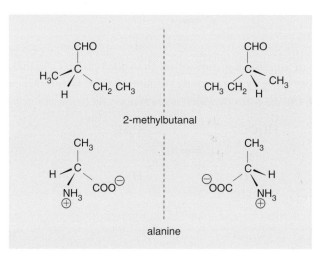

Fig. 5.1 These compounds are optical isomers – they are chiral, because they are non-superimposable mirror images.

that rotate the plane to the right (clockwise, looking into the sample) are **dextrorotatory**, and this property is shown by a (+) in front of the name. Molecules rotating the plane of polarisation of light to the left are **laevorotatory**, and this is shown by a (−).

The standard light that is used for measurements of optical activity is sodium light. The light must be monochromatic because the angle of rotation depends (among other things) on the wavelength of the light used. For some wavelengths it will be zero.

If a solution contains equal (molar) amounts of each of the two forms of a chiral molecule it is called a **racemic mixture**, and this shows no rotation. The clockwise rotation of one isomer is cancelled by the anticlockwise rotation from the other one. Most chemical processes that give rise to chiral molecules produce racemic mixtures.

Chirality is of great significance in biochemistry. Many molecules exist in organisms in only one of the two forms. Glucose is a chiral molecule (with several chiral centres); its mirror-image has no taste, and cannot even be absorbed from the gut. Carvone, a natural product, tastes (and smells) of spearmint or of caraway seeds, depending on which isomer is considered.

The classification of reaction types

An important feature of organic chemistry is the study of reaction mechanisms. Mechanisms enable the representation of the movement of electrons, and hence bonds, during a reaction. The detailed study of mechanisms is deferred until Unit 5, but it is useful here to have a list of the various reaction types, given in Table 5.2. The classification of reagents as nucleophiles or electrophiles was covered in Unit 2.

Grignard reagents

At the end of the nineteenth century, chemists such as François Grignard and Percy Frankland (Figures 5.2 and 5.3) discovered the synthetic potential of organometallic compounds. An organometallic compound is one in which a metal is linked to a carbon atom by a bond with significant covalent character.

Fig. 5.2 François Grignard (1871–1935); French chemist who discovered the Grignard reagent.

There will always be some ionic character to the metal–carbon bond because the electronegativities of metals are less than that of carbon. A most important feature of this ionic character is that the carbon, unusually, carries a partial negative charge. This means that the carbon atom can behave as a nucleophile.

Grignard found that halogenoalkanes would react with magnesium in a solvent of dry ethoxyethane (ether) when they were heated together under reflux:

$$C_2H_5I \ + \ Mg \ \xrightarrow{\text{dry ether}} \ C_2H_5MgI$$
ethyl magnesium iodide

The ether must be perfectly dry since water destroys the resulting **Grignard reagent**:

$$C_2H_5MgI \ + \ H_2O \ \rightarrow \ C_2H_6 \ + \ Mg^{2+} \ + \ OH^- \ + \ I^-$$

This is not a useful reaction of Grignard reagents; alkanes are not difficult to obtain.

A trace of iodine helps to initiate the reaction, and the choice of solvent is crucial. The ether, $C_2H_5OC_2H_5$, helps to stabilise the organometallic complex by solvating the combined magnesium, in a similar way to that in which the solvation energy of the sodium and chloride ions allows crystalline sodium chloride to dissolve in water. The Grignard reagents are always used in solution and are never isolated. They can be purchased ready-made.

Fig. 5.3 Percy Frankland (1858–1946) also helped elucidate the properties of organometallic compounds.

Table 5.2 *A summary of the reaction types covered in this Unit*

Compound	Reagent	Product	Reaction type
Grignard reagent RMgX	water	alkane RH	nucleophilic substitution
	carbon dioxide	carboxylic acid RCOOH	
	methanal HCHO	primary alcohol RCH_2OH	
	aldehydes R'CHO	secondary alcohol RCH(OH)R'	
	ketones R'COR''	tertiary alcohol RR'R''COH	
Carboxylic acids RCOOH	alcohol R'OH	ester RCOOR'	nucleophilic substitution followed by elimination
	lithium aluminium hydride $LiAlH_4$	alcohol RCH_2OH	reduction
	phosphorus pentachloride PCl_5	acid chloride RCOCl	nucleophilic substitution
	sodium carbonate Na_2CO_3 and sodium hydrogen carbonate $NaHCO_3$	sodium salt $RCOO^-$ Na^+	acid–base
Esters RCOOR'	aqueous mineral acid, e.g. HCl (aq)	alcohol R'OH and acid RCOOH	hydrolysis (equilibrium)
	aqueous sodium hydroxide	alcohol R'OH and salt $RCOO^-$ Na^+	hydrolysis (non-equilibrium)
Aldehydes RCHO or ketones RCOR'	hydrogen cyanide and potassium cyanide	cyanohydrin RCH(OH)CN or RR'C(OH)CN	nucleophilic addition
	2,4-dinitrophenylhydrazine	2,4-dinitrophenylhydrazone	nucleophilic addition followed by elimination
	sodium borohydride $NaBH_4$ or lithium aluminium hydride $LiAlH_4$	primary alcohol RCH_2OH or secondary alcohol RCH(OH)R'	reduction
Aldehydes RCHO	ammoniacal silver nitrate solution	silver mirror	reduction of the silver ion
	Fehling's solution	copper(I) oxide ppt	reduction of the copper(II) ion
Acid chlorides RCOCl	water	acid RCOOH	nucleophilic substitution
	ammonia	amide $RCONH_2$	
	alcohol R'OH	ester RCOOR'	
	amine $R'NH_2$	N-substituted amide RCONHR'	
Amines RNH_2	aqueous acid, e.g. HCl(aq)	RNH_3^+ Cl^-	acid–base
	acid chloride R'COCl	N-substituted amide R'CONHR	nucleophilic substitution
Amides $RCONH_2$	phosphorus(V) oxide P_4O_{10}	nitrile RCN	dehydration
	bromine followed by NaOH(aq)	amine RNH_2	substitution followed by rearrangement and elimination
Nitriles RCN	aqueous acid, e.g. HCl(aq)	acid RCOOH	hydrolysis
	aqueous sodium hydroxide	salt $RCOO^-$ Na^+	
	lithium aluminium hydride $LiAlH_4$	amine RCH_2NH_2	reduction
Amino acids $RCH(NH_3^+)COO^-$	aqueous acid, e.g. HCl(aq)	salt $RCH(NH_3^+)COOH$	acid–base
	aqueous sodium hydroxide	salt $RCH(NH_2)COO^-$ Na^+	

Reactions and synthetic importance of Grignard reagents

The unusual role of carbon as a carbanion makes the Grignard reagent a nucleophile. When looking at how these reagents behave, it is difficult to represent the polarity of the Mg–C bond when the link between the Mg atom and the ion of X is ionic. Structures like that in Figure 5.4a are undoubtedly more accurate than that in Figure 5.4b, because the second structure fails to show the ionic nature of the halide bond. However, the latter is the more usual representation of the structure.

Reactions with aldehydes and ketones

(a) More accurate representation   (b) The usual representation

Fig. 5.4 Charge distribution in a Grignard reagent.

Nucleophilic addition to an aldehyde or ketone and decomposition of the resulting complex with dilute acid gives secondary or tertiary alcohols, respectively:

$$
\begin{array}{ccccc}
\underset{H}{\overset{R}{\diagdown}} C = O & \rightarrow & \underset{RCH_2}{\overset{R}{\diagdown}} C \diagup^{OMgX}_{\diagdown H} & \xrightarrow{HCl(aq)} & \underset{RCH_2}{\overset{R}{\diagdown}} C \diagup^{OH}_{\diagdown H} \quad (+\ MgXCl)
\end{array}
$$

R.H₂C — MgX
δ–   δ+

2° alcohol

QUESTION
Show how tertiary alcohols are formed from ketones by a similar reaction.

The complex could be hydrolysed by water alone, but the resulting suspension of magnesium hydroxide and basic magnesium salts would be far harder to deal with than a solution. Use of methanal, HCHO, enables the production of a primary alcohol.

Reaction with carbon dioxide

The addition of a Grignard reagent to carbon dioxide is a convenient method of extending a carbon chain:

$$
O = C = O \quad \rightarrow \quad RCH_2 - C \diagup^{O}_{\diagdown OMgX} \quad \xrightarrow{HCl(aq)} \quad RCH_2 - C \diagup^{O}_{\diagdown OH} \quad (+MgXCl)
$$

R.H₂C — MgX
δ–   δ+

Apart from its use as a rather expensive general synthetic method, this reaction has proved of enormous value in preparing carboxylic acids labelled with carbon-14 (^{14}C) for use in tracer experiments. One of the most convenient forms of commercially available ^{14}C is ^{14}C-labelled sodium carbonate: this is ideal for the generation of ^{14}C-enriched carbon dioxide.

Thus it is possible to use carbon dioxide with Grignard reagents to label ethanoic acid in either or both positions (C is ^{14}C):

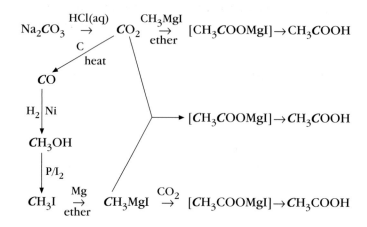

Carboxylic acids and esters

Carboxylic acids

General formula

$$C_nH_{2n+1}COOH \quad \text{or} \quad C_mH_{2m}O_2 \text{ (where } m = n+1)$$

Formulae and nomenclature

Formulae and nomenclature follow the generic name **alkanoic acid**, as shown in Table 5.3. Note that the first acid has a value of $n = 0$. This is possible since there is a carbon atom in the –COOH group.

The carboxylic acids are liquids or low-melting solids (Table 5.3). The smaller molecules are soluble in water and smell vinegary. Methanoic acid, HCOOH, is found in nettle stings and ant bites, and ethanoic acid, CH_3COOH, is found in vinegar.

Table 5.3 *Melting and boiling temperatures of carboxylic acids*

Formula	Name	Mp/°C	Bp/°C
HCOOH	methanoic acid	8	101
CH_3COOH	ethanoic acid	17	118
CH_3CH_2COOH	propanoic acid	–20	141
$CH_3CH_2CH_2COOH$	butanoic acid	–8	163

Isomerism

As well as the usual ways of forming isomers, that is by changing the position of the functional group or branching the carbon chain, carboxylic acids are isomeric with another group of compounds known as esters. This will be discussed later.

Functional group and its test

The functional group is the carboxyl group.

The hydrogen atom bonded to the oxygen reacts differently from those bonded to the carbon atoms. The highly electronegative oxygen atom withdraws electrons from this H, making the loss of this as a proton relatively easy. It is this which makes this series of compounds acidic and this property can be used to test for them.

This test shows only that the compound is acidic and it would be given by any acidic compound:

$$CO_3^{2-} + 2H^+ \rightarrow H_2O + CO_2$$
or $$HCO_3^- + H^+ \rightarrow H_2O + CO_2$$

It does not in fact prove that a carboxyl group is present. A further test for the –OH group using PCl_5 as described earlier is necessary in order to show that a carboxylic acid is present.

They owe both their low volatility and their solubility in water to hydrogen-bonding. This tends to give rise to dimers in the pure acid (Figure 5.5), and accounts for the unusually high melting point of ethanoic acid when compared with other two-carbon atom compounds.

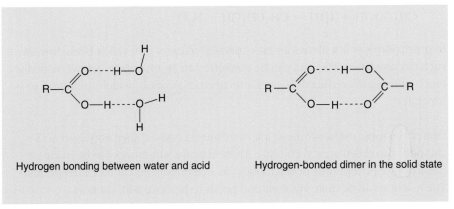

Hydrogen bonding between water and acid   Hydrogen-bonded dimer in the solid state

Fig. 5.5 Hydrogen bonding in aqueous solutions of carboxylic acids, and dimer formation.

The acids may be made by the oxidation of primary alcohols, haloform oxidation of methyl ketones, the hydrolysis of amides, cyanides and esters, and by using Grignard reagents. Aromatic carboxylic acids may also be made by the oxidation of side-chains (Topic 5.3, Unit 5).

Acidic properties of carboxylic acids

The carboxylic acids are weaker than mineral acids such as H_2SO_4, but stronger than ethanol or water, which also contain the $-OH$ group. The strength of an acid will partly depend on the stability of the anion formed compared with the original acid. If the negative charge can be delocalised around the anion, it is more likely to be formed than an anion which cannot do this. This is because, when loss of a proton (H^+) occurs, the negative charge on the oxygen is not delocalised in water and ethanol, and the electronegativity of carbon is not as great as the electronegativity of sulphur. Thus the stability of the anions decreases in the order

Carboxylic acids form the usual range of inorganic salts, those of the alkali metals and ammonia being soluble in water. The sodium and potassium salts of the higher carboxylic acids, e.g. octadecanoic (stearic) and hexadecanoic (palmitic) acids, find use as soaps, and the lithium salts are used in engineering processes as greases.

The reaction of ethanoic acid with sodium carbonate is typical:

$$2CH_3COOH(aq) + Na_2CO_3(aq) \rightarrow 2CH_3COO^- Na^+(aq) + CO_2(g) + H_2O(l)$$

Other reactions of carboxylic acids

Carboxylic acids are reduced by lithium aluminium hydride $LiAlH_4$ in dry ether solution to give primary alcohols:

$$CH_3COOH + 4[H] \rightarrow CH_3CH_2OH + H_2O$$

In practice this is not always an easy reaction to carry out, yields being low; in such circumstances the acid can be converted to an ester (see below), which is much more easily reduced. It gives a mixture of alcohols, which then have to be separated.

One of the most useful synthetic intermediates in organic chemistry is an acid chloride, also called an acyl chloride. These reactive substances can be made by the action of phosphorus pentachloride on a carboxylic acid at room temperature. The reaction can be quite vigorous and needs to be done with caution:

$$CH_3COOH + PCl_5 \rightarrow CH_3COCl + HCl + POCl_3$$
ethanoyl chloride

The reactions of acid chlorides are dealt with below.

Esters are very important compounds, used as solvents, perfumes and flavouring agents. They can be made by the reaction of a carboxylic acid with an alcohol in the presence of a small amount of concentrated sulphuric acid. The mixture is heated under reflux, the ester being produced in an equilibrium reaction:

$$CH_3COOH + CH_3CH_2OH \rightleftharpoons CH_3COOCH_2CH_3 + H_2O$$

This is not necessarily the best way to make esters; a better yield is given if the alcohol is reacted with an acid chloride, but these latter compounds are expensive and this disadvantage may offset the advantage of the esterification not being an equilibrium reaction:

$$CH_3COCl + CH_3CH_2OH \rightarrow CH_3COOCH_2CH_3 + HCl$$

The product is also cleaner since HCl is given off as a gas.

Esters
General formula

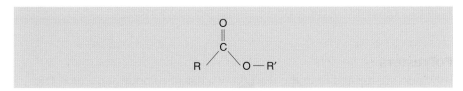

R and R′ are both alkyl groups $-C_nH_{2n+1}$. The value of n can be 0 for R but not for R′ since this would make the compound a carboxylic acid.

Formulae and nomenclature

Esters are derivatives of carboxylic acids and the stem of the name is **carboxylate**, according to the acid from which they are derived. Hence the stem depends on the nature of R. For example, ethanoic acid gives esters called ethanoates, methanoic acid gives methanoates, etc. The name of the ester is then completed by a prefix which denotes the name of the alkyl group R′. Some examples are shown in Table 5.4.

Table 5.4 *The names of some esters*

	Formula	Name
1	$CH_3CO_2CH_3$	methyl ethanoate
2	$CH_3CO_2CH_2CH_3$	ethyl ethanoate
3	$CH_3CH_2CO_2CH_3$	methyl propanoate
4	$HCO_2CH_2CH_3$	ethyl methanoate

Isomerism

There are several examples of isomerism shown within Table 5.4, for example, 1 and 4, or 2 and 3. Apart from this, however, esters are isomeric with carboxylic acids. For example, methyl ethanoate is isomeric with propanoic acid (both are $C_3H_6O_2$), as well as with ethyl methanoate.

Carboxylic acids always have a higher boiling point than the isomeric esters and also respond to the test for an acid (see above). Esters cannot respond to this test, there being no acidic hydrogen atom present.

Esters are formed by heating the acids with alcohols in the presence of a catalyst, usually concentrated sulphuric acid:

$$ROH \ + \ R'COOH \ \overset{H^+}{\rightleftharpoons} \ R'COOR \ + \ H_2O$$

As the esters can no longer have intermolecular hydrogen-bonding, they are more volatile than the acids (and possibly more volatile than the alcohols from which they were made) unless the alkyl group R is very large in comparison with the group R'. Thus the esters can often be separated from the reaction mixture by continuous distillation, which drives the equilibrium to the right.

Esters are employed in food flavourings, perfumes and cosmetics, solvents for glues, varnishes and spray-paints. The alkaline hydrolysis of esters, called 'saponification', is the basis of soap manufacture (Unit 5). Because the acid species is in the anionic form (as a salt), the alkaline hydrolysis, unlike the acid hydrolysis, is irreversible and does not reach equilibrium. Esters of propane-1,2,3-triol (glycerol) occur naturally as oils, fats and waxes.

Polyesters

Polyesters are formed from the reaction of a diacid chloride and a diol. The commonest such polymer is Terylene, which is formed from ethane-1,2-diol and the acid chloride of benzene-1,4-dicarboxylic acid (terephthalic acid):

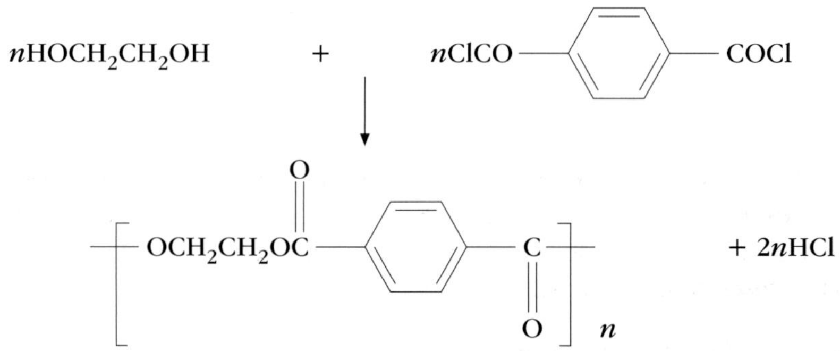

Polyesters are widely used for fabrics for clothes manufacture. They are attacked under strongly acidic or alkaline conditions since the ester bonds are hydrolysed.

Hydrolysis of esters

The term 'hydrolysis' literally means 'breaking by water'. In organic chemistry, however, reaction with water is almost invariably slow even when it is possible. A much faster reaction can be brought about by using a dilute mineral acid or a dilute alkali, but even then heating under reflux is necessary. The reaction of haloalkanes with aqueous sodium hydroxide can therefore be considered to be hydrolysis.

The hydrolysis of esters is best brought about by boiling under reflux with dilute sodium hydroxide. The reaction gives the sodium salt of a carboxylic acid, together with an alcohol. For example:

$$CH_3COOCH_2CH_3 \ + \ NaOH \ \rightarrow \ CH_3COO^- \ Na^+ \ + \ CH_3CH_2OH$$

  ethyl ethanoate   sodium ethanoate   ethanol

While acids ionise and form salts by fission of the –O–H bond, esters are formed and hydrolysed by breaking and making the adjacent C–O bond. The group RC=O is known as an acyl group, and this mechanism is accordingly known as acyl-oxygen fission.

Aldehydes and ketones

These two homologous series are usually considered together since they have many reactions in common.

Aldehydes and ketones are widely distributed in nature; for example sugars, and pyruvate CH_3COCOO^- in metabolic pathways. Their conversion to many types of alcohol and to acids makes them very useful in synthetic organic chemistry.

General formulae

$C_nH_{2n}O$

Both groups have the same general formula; to avoid confusion, they must be written so as to show their structure.

R and R′ are alkyl groups which may be the same or different. In the case of aldehydes, R can also be a hydrogen atom, but in the case of ketones both groups *must* contain at least one carbon atom. Hence the simplest ketone contains three carbon atoms.

Formulae and nomenclature

The formulae and nomenclature follow from the generic names **alkanal** and **alkanone**, as shown in Table 5.5.

Table 5.5 *Names of aldehydes and ketones*

Aldehydes	
Formula	**Name**
HCHO	methanal
CH_3CHO	ethanal
CH_3CH_2CHO	propanal
$CH_3CH_2CH_2CHO$	butanal
$CH_3CH(CHO)CH_3$	2-methylpropanal

Ketones	
Formula	**Name**
CH_3COCH_3	propanone
$CH_3CH_2COCH_3$	butanone
$CH_3CH_2CH_2COCH_3$	pentan-2-one
$CH_3CH_2COCH_2CH_3$	pentan-3-one

TEST

Test for the carbonyl group
This test is for both aldehydes and ketones.
- Reagent. Add an excess of a solution of 2,4-dinitro-phenylhydrazine.
- Result. An orange-yellow precipitate is formed.

TEST

Test for the –CHO group
There are two possible tests, either of which can be used. These tests are given by aldehydes only.

Test 1
- Reagent. Add an ammoniacal solution of silver nitrate and warm. (This solution is made by adding a few drops of dilute sodium hydroxide to a solution of silver nitrate followed by dilute ammonia solution until the brown precipitate dissolves.)
- Result. Silver metal is precipitated (often in the form of a mirror on the side of the test tube although it may also appear as a black precipitate).

Test 2
- Reagent. Add Fehling's or Benedict's solution and warm.
- Result. A red precipitate (of copper(I) oxide) is formed.

There is no alkyl group in methanal, but there is a carbon atom in the functional group.

Isomerism

Isomerism can occur *within* each homologous series, as can be seen in the examples; butanal is isomeric with 2-methylpropanal, and pentan-2-one is isomeric with pentan-3-one. Isomers *within* a homologous series have similar chemical properties but different physical properties (such as boiling temperatures). Isomers *between* homologous series have different chemical and physical properties.

Isomerism can also occur *between* the two series for molecules with at least three carbon atoms, that is aldehydes are isomeric with ketones. Thus propanal is isomeric with propanone, butanal with butanone (as well as with 2-methylpropanal), etc. In such cases, the isomers will differ in chemical reactions as well as physical properties. A method of distinguishing between the two series is shown in the following section.

Functional groups and tests

The functional groups are –CHO for aldehydes and $>$C=O for ketones. Both groups contain a $>$C=O or carbonyl group.

Both tests for the –CHO group depend on the fact that aldehydes are good reducing agents, while ketones show no reducing properties. The aldehyde is oxidised to a carboxylic acid in both cases:

$$-CHO + [O] \rightarrow -COOH$$

With the exception of methanal (HCHO), which is a gas at room temperature, the lower members are volatile liquids (Table 5.6) that are soluble in water. The ketones usually smell more pleasant than the aldehydes. The polarity of the carbonyl group makes the compounds less volatile than alkanes of similar size, but they are more volatile than the alcohols, which are hydrogen-bonded. The solubility of the smaller carbonyl compounds in water is the result both of hydrogen-bonding and addition of the water at the carbonyl group (Figure 5.6).

With rare exceptions (for example that of CCl_3CHO), the hydrates are too unstable to be isolated.

Table 5.6 *Boiling temperatures of aliphatic aldehydes*

Formula	Name	Bp/°C
HCHO	methanal	–21
CH_3CHO	ethanal	21
C_2H_5CHO	propanal	49
C_3H_7CHO	butanal	75

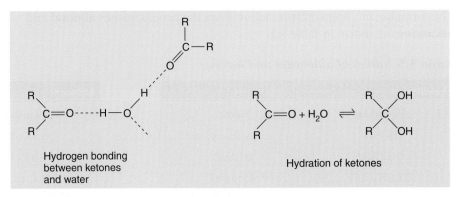

Hydrogen bonding between ketones and water

Hydration of ketones

Fig. 5.6 Hydrogen bonding and hydration in ketones.

Oxidation and reduction

You have already met the oxidation of aldehydes, both as intermediates in the oxidation of primary alcohols (Unit 1) and in their capacity to reduce Fehling's or Benedict's solution and ammoniacal silver nitrate (above).

Normally, ketones are not easy to oxidise, but methyl ketones (-2-ones) can be oxidised by iodine under alkaline conditions (sodium hydroxide solution) with loss of the methyl carbon atom; the product is iodoform, CHI_3, and the sodium salt of the carboxylic acid (having one fewer carbon atoms).

While this 'haloform reaction' has been used to degrade methyl ketones to acids, it is more useful as a test of structure. The test is performed by warming the suspected methyl ketone with iodine and sodium hydroxide solution (or sodium chlorate(I) and an iodide) and, when positive, gives a yellow precipitate of triiodomethane, CHI_3 (iodoform). Care must be exercised in its interpretation, since it gives a positive test not only with compounds having the structure CH_3CO- but also with compounds with the structure $CH_3CH(OH)-$, which becomes CH_3CO- in the oxidising conditions employed. **The iodoform reaction is *not* a general test for ketones**.

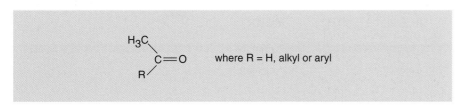

H_3C, $C=O$ R where R = H, alkyl or aryl

Fig. 5.7 Structure of a methyl ketone.

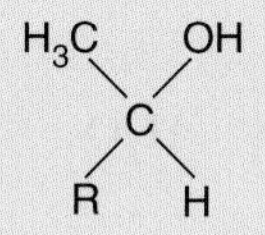

H_3C OH C R H where R = alkyl or aryl

Fig. 5.8 Structure of a methyl secondary alcohol.

The structures in Figures 5.7 and 5.8 represent methyl ketones or methyl secondary alcohols. If R = H, when they are the special cases of ethanal and ethanol, respectively. It is thus very important not to use ethanol as a solvent for this reaction.

The reaction may be represented by a variety of equations, but the easiest is shown below for butan-2-ol. If necessary, as here, oxidation first occurs to give a carbonyl compound:

$$CH_3CH(OH)C_2H_5 + I_2 \rightarrow CH_3COC_2H_5 + 2HI$$

Substitution then occurs in the methyl group:

$$CH_3COC_2H_5 + 3I_2 + 3NaOH \rightarrow CI_3COC_2H_5 + 3NaI + 3H_2O$$
$$\text{1,1,1-triiodobutan-2-one}$$

$$CI_3COC_2H_5 + OH^- \rightarrow CHI_3 + CH_3CH_2COO^-$$

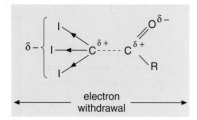

Fig. 5.9 Lowering of electron density by the triiodomethyl group in the bond to be hydrolysed.

The electronegativity of the three halogen atoms exerts a strong electron-withdrawing effect on the adjacent C–C bond. The carbonyl group is already withdrawing electrons in the other direction. This lowers the electron density in the C–C bond joining the CI$_3$ to the carbonyl group (Figure 5.9), and nucleophilic attack on the carbonyl carbon is easily followed by fission of this bond.

Reductions by lithium aluminium hydride in ether

Lithium aluminium hydride, LiAlH$_4$ (also named lithium tetrahydridoaluminate(III)), in dry ethoxyethane (ether) is capable of reducing all compounds containing the C=O (carbonyl) group. The reagent is expensive and is not easy to store because it rapidly decomposes in damp conditions. It is not suitable for industrial reductions except for small-scale use in the drugs industry since it is very expensive. The following are examples of reductions of various carbonyl compounds by lithium aluminium hydride; the reducing power is shown by [H]:

$$RCHO \quad + \quad 2[H] \quad \rightarrow \quad RCH_2OH$$

aldehyde   1° alcohol

$$RCOR' \quad + \quad 2[H] \quad \rightarrow \quad \underset{\underset{OH}{|}}{RCHR'}$$

ketone   2° alcohol

$$RCOOR' \quad + \quad 4[H] \quad \rightarrow \quad RCH_2OH \quad + \quad R'OH$$

ester

$$RCOOH \quad + \quad 4[H] \quad \rightarrow \quad RCH_2OH \quad + \quad H_2O$$

acid

In each reaction, a complex is formed with the aluminium, which is stabilised by the dry ether (cf. Grignard reagents). The mechanisms are rather long and complicated because they involve the successive steps in the substitution of the AlH$_4^-$ ion to give the complex. Reduction may be considered to begin by the nucleophilic addition of the hydride, H$^-$, ion from the AlH$_4^-$ ion at the carbon atom of the carbonyl group:

$$R'-\underset{\underset{H-AlH_3^-}{\curvearrowleft}}{\overset{\overset{R}{|}}{C}}=O \rightarrow R-\underset{\underset{H}{|}}{\overset{\overset{R}{|}}{C}}-OAlH_3^-$$

The complex is then decomposed by water to release the alcohol product.

Reduction of carbonyl compounds can be brought about by less expensive reagents than lithium aluminium hydride, such as catalytic hydrogenation or the slightly less powerful sodium borohydride, $NaBH_4$. Sodium borohydride (sodium tetrahydridoborate(III)) has the advantage of being usable in aqueous solution. Thus there is no need for the rigorously anhydrous conditions demanded by $LiAlH_4$. However, $NaBH_4$ will reduce only aldehydes and ketones and not acids or esters.

The initial addition of H^- by $LiAlH_4$ is a nucleophilic attack and suited to reaction with the positively charged carbon of the carbonyl group. Alkenes are susceptible to electrophilic attack and are not affected by $LiAlH_4$. Thus a compound such as pent-4-en-2-one would be totally reduced by catalytic hydrogenation with hydrogen and platinum, but would retain the $C{=}C$ double bond if reduced by $LiAlH_4$:

$$H_2C{=}CHCH_2COCH_3 \xrightarrow{H_2/Pt} H_3CCH_2CH_2CH(OH)CH_3$$
$$\text{pent-4-en-2-one}$$

$$H_2C{=}CHCH_2COCH_3 \xrightarrow{LiAlH_4/ether} H_2C{=}CHCH_2CH(OH)CH_3$$

$LiAlH_4$ is therefore a useful selective reducing agent.

Other nucleophilic reactions at the carbonyl group

Both aldehydes and ketones undergo a wide range of addition and addition–elimination (condensation) reactions. Two will be discussed here: one is of synthetic and one of analytical importance.

Nucleophilic addition of cyanide

Hydrogen cyanide is covalent and a weak acid. To provide the cyanide ion, CN^-, an aqueous solution of sodium or potassium cyanide is used:

$$RCHO \ + \ HCN \ \rightleftharpoons \ RCH{\Big\langle}{\overset{\text{OH}}{\underset{\text{CN}}{}}}$$

Aqueous hydrogen cyanide alone is not a suitable reagent since it is a weak acid, and its conjugate base, the cyanide ion, is strong and will remove a proton from water:

$$:CN^- \ + \ H_2O \ \rightleftharpoons \ HCN \ + \ :OH^-$$

Using hydrogen cyanide in the mixture will drive the equilibrium to the left. In practice KCN in aqueous ethanol at about pH 8 is used.

The resulting cyanohydrins are of synthetic value, as acid hydrolysis (boiling with dilute HCl) gives α-hydroxyacids:

$$CH_3C{\Big\langle}{\overset{\text{OH}}{\underset{\text{CN}}{}}}{-}H \ \xrightarrow{H_2O/H^+} \ CH_3{-}\underset{\underset{\text{COOH}}{|}}{\overset{\overset{\text{OH}}{|}}{C}}{-}H$$

$$\text{2– hydroxypropanoic acid}$$
$$\text{(lactic acid)}$$

Nucleophilic addition–elimination with 2,4-dinitrophenylhydrazine

Carbonyl compounds undergo a wide range of reactions with nucleophiles containing nitrogen. In the nineteenth century, hydrazine (H_2NNH_2) was found to give crystalline derivatives (hydrazones) with carbonyl compounds, but these derivatives were unsuitable for the identification of the carbonyl compound because their melting points were often indefinite. Their reactions were complicated because either or both nitrogen atoms could react. The use of phenylhydrazine ($C_6H_5NHNH_2$) and its 2,4-dinitro derivative, however, was very successful.

2,4-Dinitrophenylhydrazine is particularly useful because the derivatives it gives with carbonyl compounds have low solubility in many solvents, and the formation of an orange precipitate can be used as a simple test for the presence of a carbonyl group.

In addition, their melting temperatures are well-documented and can be used to identify the aldehyde or ketone originally used.

$$
\begin{array}{l}
R \\
\diagdown \\
C{=}O \ + \ H_2NNHAr \ \rightarrow \\
\diagup \\
R'
\end{array}
\qquad
\begin{array}{l}
R OH \\
\diagdown\diagup \\
C \\
\diagup\diagdown \\
R' NHNHAr
\end{array}
$$

$$
\begin{array}{l}
R OH \\
\diagdown\diagup \\
C \rightarrow \\
\diagup\diagdown \\
R' NHNHAr
\end{array}
\qquad
\begin{array}{l}
R \\
\diagdown \\
C{=}NNHAr \\
\diagup \\
R' \\
+H_2O
\end{array}
$$

e.g. R = R' = CH_3

and

$$Ar = \text{—} \overset{\displaystyle }{\underset{O_2N}{\bigcirc}} NO_2$$

> ## QUESTION
>
> Why should replacing OH with Cl make acid chlorides more reactive than acids towards nucleophiles?

Acid chlorides (acyl chlorides)

Reaction of a carboxylic acid produces, as we have already seen, an acid chloride:

$$CH_3COOH \ + \ PCl_5 \ \rightarrow \ CH_3COCl \ + \ HCl \ + \ POCl_3$$
ethanoyl chloride

Ethanoyl chloride is a fuming, colourless liquid.

Acid chlorides are more reactive in nucleophilic substitution reactions than the halogenoalkanes such as CH_3CH_2Cl. This is because the carbon atom of ethanoyl chloride bears two electron-withdrawing groups rather than just one, and so is more δ^+ than the carbon atom in chloroethane. It is also easier to attack from a stereochemical point of view, since the ethanoyl chloride molecule is flat whereas chloroethane is tetrahedral about the attacked carbon atom. The reaction of ethanoyl chloride with water can be violent, and furthermore is useless since it converts ethanoyl chloride (expensive) into the cheap compound from which it was made:

$$CH_3COCl \ + \ H_2O \ \rightarrow \ CH_3COOH \ + \ HCl$$

This reaction causes the liquid to fume in air, and is responsible for its choking smell. Other acid chlorides are less volatile and less easily hydrolysed, but often cause stinging, watering eyes – they are *lachrymatory*. Benzoyl chloride C_6H_5COCl is one such material.

All of the reactions given below occur at room temperature.

Acyl halides form:

esters with alcohols,

$$CH_3COCl \ + \ C_2H_5OH \ \rightarrow \ CH_3COOCH_2CH_3 \ + \ HCl$$
ethyl ethanoate

amides with ammonia,

$$CH_3COCl \ + \ 2NH_3 \ \rightarrow \ CH_3CONH_2 \ + \ NH_4Cl$$
ethanamide

and N-substituted amides with primary amines,

$$CH_3COCl \ + \ C_6H_5NH_2 \ \rightarrow \ CH_3CONHC_6H_5 \ + \ HCl$$
phenylamine   *N*-phenylethanamide

Amines

Primary amines are compounds having an $-NH_2$ group as the functional group. There are related compounds, secondary ($RR'NH$) and tertiary ($RR'R''N$) amines, but these are not considered further here.

Methylamine is a gas at room temperature, but the majority of the lower amines are liquids (Table 5.7). They are less volatile than the corresponding alkanes (which are attracted only by the weakest van der Waals' forces) or the halogenoalkanes (which have permanent dipoles) because the amines are hydrogen-bonded (Figure 5.10). However, the lower amines are more volatile than the alcohols or the amides, which are more strongly or extensively hydrogen-bonded.

The amines have unpleasant smells, which change from ammoniacal to fishy as the molecular mass increases. The smaller amines are soluble in water because hydrogen-bonding occurs between the amino group and the water (Figure 5.10).

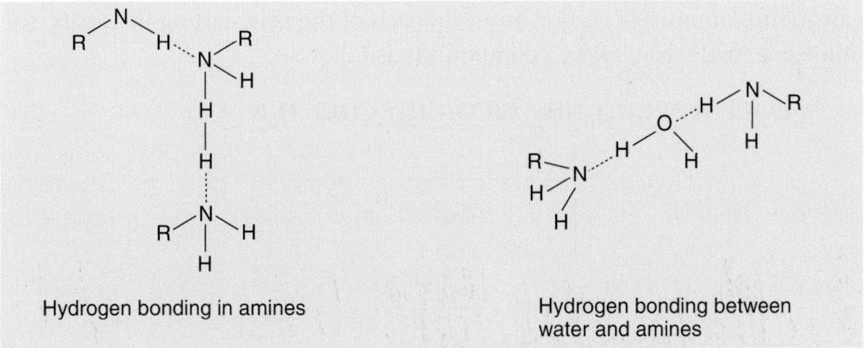

Table 5.7 *Boiling temperatures of aliphatic primary amines*

Formula	Name	Bp/°C
CH_3NH_2	methylamine	−6
$C_2H_5NH_2$	ethylamine	17
$C_3H_7NH_2$	propylamine	49
$C_4H_9NH_2$	butylamine	76

Hydrogen bonding in amines   Hydrogen bonding between water and amines

Fig. 5.10 Hydrogen bonding in amines, and between amines and water.

All amines are basic. The aliphatic amines are slightly more basic than ammonia because of the small electron-donating effect of the alkyl group, but the pH of saturated aqueous solutions falls as the molecular mass increases because of decreasing solubility. Aromatic amines are much less basic and much less soluble in water. The lone pair of electrons on the nitrogen atom is responsible for the basic properties.

All the amines form water-soluble salts with mineral acids (HCl, H_2SO_4, etc.):

$$CH_3CH_2NH_2 + HCl \rightleftharpoons CH_3CH_2\overset{+}{N}H_3Cl^-$$

$$\underset{\substack{\text{phenylamine} \\ \text{(aniline)}}}{C_6H_5NH_2} + HCl \rightleftharpoons \underset{\substack{\text{phenylammonium chloride} \\ \text{(aniline hydrochloride)}}}{C_6H_5\overset{+}{N}H_3Cl^-}$$

Thus amines that do not dissolve in water form solutions in acids and are reprecipitated by alkalis:

$$\underset{\substack{\text{sparingly} \\ \text{soluble}}}{C_6H_5NH_2(l)} \underset{OH^-(aq)}{\overset{H^+(aq)}{\rightleftharpoons}} \underset{\substack{\text{appreciably} \\ \text{soluble}}}{C_6H_5\overset{+}{N}H_3(aq)}$$

Amines also react with acid chlorides since they are nucleophiles. This reaction has already been mentioned above:

$$C_6H_5NH_2 + CH_3COCl \rightleftharpoons \underset{N\text{-phenylethanamide}}{C_6H_5NHCOCH_3} + HCl$$

These substituted amides are sometimes used as pharmaceuticals, e.g. paracetamol.

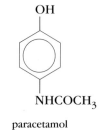

paracetamol

Reaction with acid chlorides

One of the most important reactions of amines is with acid chlorides to form polyamides. The first such compound was nylon, first synthesised by Wallace Carothers in the 1940s.

Polyamides are condensation polymers, i.e. compounds formed between monomers that react by the elimination of a small molecule. The various forms of nylon can made by the reaction of a diacid chloride with a diamine, so the small molecule that is eliminated is HCl. The different forms are referred to by means of the number of carbon atoms in each of the two starting materials, so Nylon-6,6 is made from two C_6 compounds as follows:

$$-COCl \quad H_2N(CH_2)_6NH_2 \quad ClCO(CH_2)_4COCl \quad H_2N-$$

$$\downarrow$$

$$-\left(\overset{O}{\underset{\underset{H}{|}}{\overset{\|}{C}}}-N(CH_2)_6N-\overset{O}{\overset{\|}{C}}(CH_2)_4\overset{O}{\overset{\|}{C}}-\overset{H}{\overset{|}{N}}\right)-$$

$$\underset{H}{\qquad}$$

HCl   HCl   HCl

Nylon is a tough, hard polymer with high impact resistance and low friction, and among many other uses finds application in the manufacture of precision gearing. The stiffness of the polymer is a consequence of strong hydrogen bonding between adjacent fibre chains. Millions of tonnes of nylon are made each year.

Amides

Amides are acyl derivatives of ammonia. With the exception of methanamide, they are solids (Table 5.8), and the lower members are all soluble in water. Both these properties are caused by hydrogen-bonding (Figure 5.11). The ability to form dimers gives the amides their relatively high melting temperatures.

As we have seen, they can be made by the action of ammonia on acyl halides:

$$CH_3COCl + 2NH_3 \rightarrow CH_3CONH_2 + NH_4Cl$$

Unlike amines, amides are not appreciably basic: their aqueous solutions are neutral. The lone pair of electrons on the nitrogen atom, which is responsible for basic character, is made less available by interaction with the π-system of the carbonyl group (Figure 5.12). At the same time, the effect of the oxygen atom in creating a positive charge on the C atom of the carbonyl group is reduced. Thus amides, like acids and esters, do not undergo the nucleophilic addition reactions of aldehydes and ketones with, for example, cyanide ions.

Hofmann degradation of amides

Amides have little synthetic value as intermediates, but sometimes they can be used as a means of introducing the amino group $-NH_2$ with the simultaneous loss of a carbon atom. This reaction is known as the Hofmann degradation: a degradation implies the loss of a carbon atom from a molecule. It is brought about by the action of liquid bromine followed by sodium hydroxide solution and heat:

$$RCONH_2 + Br_2 + 4NaOH \rightarrow RNH_2 + Na_2CO_3 + 2NaBr + 2H_2O$$

Note the loss of the carbon atom.

Dehydration

Amides can be dehydrated by heating with phosphorus(v) oxide, P_4O_{10}; this is a way of preparing nitriles (see below):

$$6RCONH_2 + P_4O_{10} \rightarrow 6RCN + 4H_3PO_4$$

Thus it can be seen that an amide can be converted into an amine with either the same number of carbon atoms or one fewer:

$$
\begin{array}{cccccc}
& \text{NaOH/Br}_2 & & \text{P}_4\text{O}_{10} & & \text{H}_2\text{/Ni} \\
RNH_2 & \leftarrow & RCONH_2 & \rightarrow & RCN & \rightarrow & RCH_2NH_2 \\
& & \text{amide} & -\text{H}_2\text{O} & \text{or other} \\
& & & & \text{reducing agent e.g. LiAlH}_4
\end{array}
$$

Table 5.8 *Melting temperatures of aliphatic amides*

Formula	Name	Mp/ °C
$HCONH_2$	methanamide	3
CH_3CONH_2	ethanamide	82
$C_2H_5CONH_2$	propanamide	81
C_3H_7CONH	butanamide	115

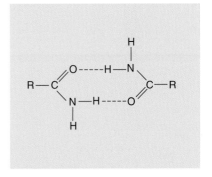

Fig. 5.11 Intermolecular hydrogen bonding in amides.

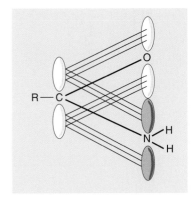

Fig. 5.12 Orbital overlap in amides.

Nitriles

Nitriles are organic cyanides, and have the functional group –CN. Ethanenitrile CH_3CN, also called methyl cyanide, is the first member of the series. Their chemistry principally involves their hydrolysis to carboxylic acids (or their salts) and their reduction to amines.

Hydrolysis of nitriles can be effected by heating them with an aqueous solution of either hydrochloric acid or sodium hydroxide. The first gives the free acid, the second its salt; hydrolysis proceeds via the amide as intermediate, but under the conditions used this is always hydrolysed itself:

$$CH_3CH_2CN + HCl + 2H_2O \rightarrow CH_3CH_2COOH + NH_4Cl$$

$$CH_3CH_2CN + NaOH + H_2O \rightarrow CH_3CH_2COONa + NH_3$$

The reduction of nitriles can be performed with lithium aluminium hydride in dry ether, followed by cautious treatment with water:

$$CH_3CH_2CN + 4[H] \rightarrow CH_3CH_2CH_2NH_2$$

The use of a nitrile can therefore enable an increase of one carbon atom in the length of a carbon chain:

$$CH_3CH_2Br \rightarrow CH_3CH_2CN \rightarrow CH_3CH_2CH_2NH_2$$

Amino acids

Amino acids – more accurately α-amino acids – are compounds nominally of the form:

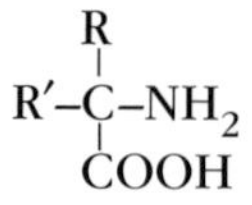

$$
\begin{array}{c}
R \\
| \\
R'-C-NH_2 \\
| \\
COOH
\end{array}
$$

but in fact they always exist in the zwitterion form shown below.

The simplest is glycine, where R = R' = H. There are some 20 amino acids that make up proteins, and all but glycine are chiral.

Amino acids possess both an acidic (–COOH) and a basic (–NH$_2$) functional group. They are solids, with melting temperatures that are quite high for such small molecules; glycine, for example, melts with decomposition at 262°C. The reason for this is that the molecules have ionic interactions, because the carboxyl group hydrogen atom has protonated the amino group nitrogen atom to give a *zwitterion*, an ion having both a positive and a negative charge:

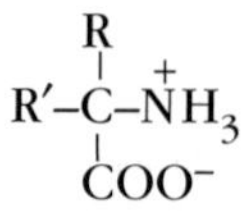

$$
\begin{array}{c}
R \\
| \\
R'-C-\overset{+}{N}H_3 \\
| \\
COO^-
\end{array}
$$

The interactions between the zwitterions are therefore strong.

> ### QUESTION
>
> Suggest a powerful reducing agent that might be able to reduce an amide directly to the amine with the same number of carbon atoms.

Amino acids react with both acids and bases; they are amphoteric. Thus:

$$^-OOCCH_2\overset{+}{N}H_3 + OH^- \rightarrow\ ^-OOCCH_2NH_2 + H_2O$$

$$^-OOCCH_2\overset{+}{N}H_3 + H^+ \rightarrow HOOCCH_2\overset{+}{N}H_3$$

This is sometimes quoted as evidence for the buffering action of amino acids. To be a buffer, however, both the acid and its conjugate base ($HOOCCH_2\overset{+}{N}H_3$ and $^-OOCCH_2\overset{+}{N}H_3$) or the base and its conjugate acid ($^-OOCCH_2\overset{+}{N}H_3$ and $^-OOCCH_2NH_2$) must be present. So once some acid or some base has been added, the mixture of species present will indeed act as a buffer, but the pH will be very different from the solution of the amino acid on its own.

Assessment questions

The following questions are all taken from Edexcel Unit 4 tests for the years 2002 and 2003.

1 (a) Write an equation which represents the change when the second electron affinity of oxygen is measured. **[2]**

(b) Construct a Born–Haber cycle and use it and the data below to calculate the second electron affinity of oxygen.

	$\Delta H/\text{kJ mol}^{-1}$
Enthalpy of atomisation of magnesium	+150
Bond energy of O=O in oxygen	+496
1st ionisation energy of magnesium	+736
2nd ionisation energy of magnesium	+1450
1st electron affinity of oxygen	−142
Lattice enthalpy of magnesium oxide	−3889
Enthalpy of formation of magnesium oxide	−602

[4]

(c) (i) MgO(s) has the same crystal structure as NaCl(s). The lattice enthalpy of NaCl(s) is −771 kJ mol^{-1} whilst that of MgO(s) is −3889 kJ mol^{-1}.

Explain the difference in lattice enthalpies. **[4]**

(ii) Despite its high lattice enthalpy sodium chloride is soluble in water.

What other factor is important in enabling compounds such as sodium chloride to be soluble in polar solvents such as water? **[1]**

(iii) Explain why magnesium oxide is insoluble in water. **[2]**

(Total 13 marks)
(June 2002)

2 (a) From the compounds of the elements in Group 4 of the Periodic Table, carbon to lead, give the **formula** of:

(i) an acidic oxide;

(ii) an oxide which can behave as a base. **[2]**

(b) Aluminium oxide is an amphoteric oxide.

(i) What is meant by the term **amphoteric?** **[1]**

(ii) Write two **ionic** equations which show the amphoteric behaviour of aluminium oxide, including state symbols. **[4]**

(c) (i) Complete the table below by writing balanced equations, without state symbols, for the reactions of the species with water. In each case suggest the approximate pH of the solution formed by adding about 1 g of the substance to 100 cm^3 of water. **No calculations are necessary**.

	Equation	pH
Sodium chloride, NaCl		
Phosphorus pentachloride, PCl$_5$		

[4]

(ii) Interpret the reactions above in terms of the bonding in the compounds. **[2]**

(d) (i) Draw the shape of the molecule of carbon tetrachloride, CCl$_4$. **[1]**

(ii) Describe what happens when carbon tetrachloride is added to water. **[1]**

(iii) Describe what happens when silicon tetrachloride is added to water. **[2]**

(iv) Explain why carbon tetrachloride and silicon tetrachloride behave in different ways when added to water. **[4]**

(Total 21 marks)
(January 2002)

ASSESSMENT QUESTIONS

3 (a) Define the following terms.

   (i) pH [1]

   (ii) K_w [1]

(b) Explain the meaning of the term **strong**, as applied to an acid or a base. [1]

(c) Calculate the pH of the following solutions.

   (i) HCl(aq) of concentration 0.200 mol dm^{-3}. [1]

   (ii) NaOH(aq) of concentration 0.800 mol dm^{-3}
   ($K_w = 1.00 \times 10^{-14}$ mol^2 dm^{-6}). [2]

(d) HA is a weak acid with a dissociation constant $K_a = 5.62 \times 10^{-5}$ mol dm^{-3}).

   (i) Write an expression for the dissociation constant, K_a, of HA. [1]

   (ii) Calculate the pH of a 0.400 mol dm^{-3} solution of HA. [3]

(e) A buffer solution contains HA(aq) at a concentration of 0.300 mol dm^{-3}, and its sodium salt, NaA, at a concentration of 0.600 mol dm^{-3}. Calculate the pH of this buffer solution. [3]

(Total 13 marks)
(January 2003)

4 Consider the following equation:

$$2SO_2 + O_2 \quad 2SO_3$$

2.0 moles of SO$_2$ and 1.0 mole of O$_2$ were allowed to react in a vessel of volume 60 dm^3. At equilibrium 1.8 moles of SO$_3$ had formed and the pressure in the flask was 2 atm.

(a) (i) Write the expression for K_c for this reaction between SO$_2$ and O$_2$. [1]

   (ii) Calculate the value of K_c, with units [3]

(b) The reaction between SO$_2$ and O$_2$ is exothermic. State the effect on the following, if the experiment is repeated at a higher temperature:

   (i) K_c [1]

   (ii) the equilibrium position [1]

(c) State the effect of a catalyst on:

   (i) K_c [1]

   (ii) the equilibrium position [1]

(d) (i) Write the expression for K_p for the reaction between SO$_2$ and O$_2$. [1]

   (ii) Calculate the mole fractions of SO$_2$, O$_2$ and SO$_3$ at equilibrium. [2]

   (iii) Calculate the partial pressures of SO$_2$, O$_2$ and SO$_3$ at equilibrium. [1]

   (iv) Calculate the value of K_p, with units. [2]

(Total 14 marks)
(January 2003)

5 (a) Write the structural formulae of the organic products obtained when ethanoyl chloride reacts with the following compounds. Give the names of these products.

   (i) Ammonia, NH$_3$. [2]

   (ii) Methanol, CH$_3$OH. [2]

(b) Bromoethane reacts with magnesium to form the Grignard reagent CH$_3$CH$_2$MgBr.

This Grignard reagent reacts with:

   • CO$_2$, following by hydrochloric acid, to form compound **A**;
   • water to form compound **B**;
   • methanal, followed by hydrochloric acid, to form compound **C**.

Compounds **A** and **C** react together, in the presence of a suitable catalyst, to form compound **D**.

   (i) Write the structural formulae of compounds **A**, **B**, and **C**. [3]

   (ii) Draw the full structural formula of compound **D**. [2]

   (iii) Give the names of compounds **C** and **D**. [2]

   (iv) Identify a catalyst for the reaction between compounds **A** and **C**. [1]

(Total 12 marks)
(January 2003)

ASSESSMENT QUESTIONS

6 Consider the following reaction scheme starting from propanone.

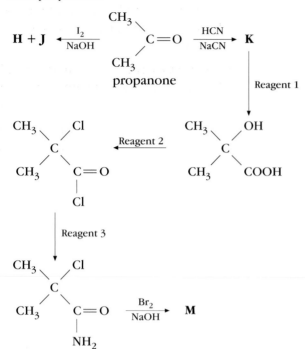

(a) Give the structural formula of **H**, **J**, **K** and **M**.
[4]

(b) Identify:
Reagent 1, Reagent 2, and Reagent 3. **[3]**

(c) Compounds produced when glucose, $C_6H_{12}O_6$, is metabolised include:

$CH_2(OH)CH(OH)CHO$ $CH_3COCOOH$
2,3-dihydroxypropanal   2-oxopropanoic acid

$CH_3CH(OH)COOH$
2-hydroxypropanoic acid

  (i) Draw the full structural formula for 2,3-dihydroxypropanal. **[1]**

  (ii) Suggest **two** of these compounds which would give a positive test with 2,4-dinitrophenylhydrazine solution. State what you would see for a positive test result. **[3]**

  (iii) Describe a test which would enable you to distinguish between the **two** compounds identified in part (ii). **[2]**

(Total 13 marks)
(June 2002)

7 *(a)* (i) Give the structural formula of a nitrile, C_4H_7N, that has an unbranched chain. **[1]**

  (ii) Primary amines can be made by reducing nitriles. Suggest a reagent that could be used for this purpose. **[1]**

  (iii) Draw the structural formula of the amine produced by reducing the nitrile given in *(a)*(i) **[1]**

(b) Draw the structure of an isomer of $C_4H_{11}N$ which has a chiral centre in the molecule and identify the chiral centre. **[2]**

(c) (i) What feature of an amine molecule makes it both a base and a nucleophile? **[1]**

  (ii) Give, by writing an equation, an example of an amine acting as a base. **[1]**

(d) Ethanoyl chloride, CH_3COCl, reacts with both amines and alcohols.

  (i) Give the name of the type of compound produced when ethanoyl chloride reacts with ethylamine, $C_2H_5NH_2$. **[1]**

  (ii) State **one** of the advantages of reacting ethanoyl chloride with ethanol to make an ester rather than reacting ethanoic acid with ethanol. **[1]**

(e) Ethanoyl chloride can be made from ethanoic acid.

  (i) Suggest a reagent suitable for this conversion. **[1]**

  (ii) Suggest how chloromethane can be converted into ethanoic acid via a Grignard reagent. (Details of the experimental apparatus are not required.) **[4]**

(Total 14 marks)
(January 2002)

8 *(a)* An alcohol with the molecular formula C_4H_9OH is chiral.

  (i) Explain what is meant by the term **chiral**. **[2]**

  (ii) Draw two diagrams to clearly represent the optical isomers that result from the chirality of this alcohol. **[2]**

(iii) Explain how you could distinguish between these two isomers experimentally. **[2]**

(b) Alcohols react with carboxylic acids to form esters.

(i) Write an equation for a typical esterification reaction. **[1]**

(ii) Suggest how this type of reaction could be used to form polyesters. Experimental details are **not** required. **[3]**

(iii) Suggest, with reasoning, whether a laboratory coat made from a polyester might be damaged by a spillage on it of hot concentrated aqueous sodium hydroxide solution. **[2]**

(iv) Give another **type** of reagent that could be used to make an ester from an alcohol. **[1]**

(Total 13 marks)
(January 2002)

The Periodic Table of Elements

Group

Period	1	2													3	4	5	6	7	0
1	1 H Hydrogen 1																			2 He Helium 4
2	3 Li Lithium 7	4 Be Beryllium 9													5 B Boron 11	6 C Carbon 12	7 N Nitrogen 14	8 O Oxygen 16	9 F Fluorine 19	10 Ne Neon 20
3	11 Na Sodium 23	12 Mg Magnesium 24													13 Al Aluminium 27	14 Si Silicon 28	15 P Phosphorus 31	16 S Sulphur 32	17 Cl Chlorine 35.5	18 Ar Argon 40
4	19 K Potassium 39	20 Ca Calcium 40	21 Sc Scandium 45	22 Ti Titanium 48	23 V Vanadium 51	24 Cr Chromium 52	25 Mn Manganese 55	26 Fe Iron 56	27 Co Cobalt 59	28 Ni Nickel 59	29 Cu Copper 63.5	30 Zn Zinc 65.4			31 Ga Gallium 70.4	32 Ge Germanium 73	33 As Arsenic 75	34 Se Selenium 79	35 Br Bromine 80	36 Kr Krypton 84
5	37 Rb Rubidium 85	38 Sr Strontium 88	39 Y Yttrium 89	40 Zr Zirconium 91	41 Nb Niobium 93	42 Mo Molybdenum 96	43 Tc Technetium (99)	44 Ru Ruthenium 101	45 Rh Rhodium 103	46 Pd Palladium 106	47 Ag Silver 108	48 Cd Cadmium 112			49 In Indium 115	50 Sn Tin 119	51 Sb Antimony 122	52 Te Tellurium 128	53 I Iodine 127	54 Xe Xenon 131
6	55 Cs Caesium 133	56 Ba Barium 137	57 La ▲ Lanthanum 139	72 Hf Hafnium 178	73 Ta Tantalum 181	74 W Tungsten 184	75 Re Rhenium 186	76 Os Osmium 190	77 Ir Iridium 192	78 Pt Platinum 195	79 Au Gold 197	80 Hg Mercury 201			81 Tl Thallium 204	82 Pb Lead 207	83 Bi Bismuth 209	84 Po Polonium (210)	85 At Astatine (210)	86 Rn Radon (222)
7	87 Fr Francium (223)	88 Ra Radium (226)	89 ▲▲ Ac Actinium (227)	104 Rf Rutherfordium (261)	105 Db Dubnium (262)	106 Sg Seaborgium (263)	107 Bh Bohrium (264)	108 Hs Hassium (269)	109 Mt Meitnerium (268)	110 Uun Ununnilium (269)	111 Uuu Unununium (272)	112 Uub Ununbium (277)								

Key

Atomic number
Symbol
Name
Molar mass in g mol⁻¹

▲ **Lanthanide elements**

58 Ce Cerium 140	59 Pr Praseodymium 141	60 Nd Neodymium 144	61 Pm Promethium (147)	62 Sm Samarium 150	63 Eu Europium 152	64 Gd Gadolinium 157	65 Tb Terbium 159	66 Dy Dysprosium 163	67 Ho Holmium 165	68 Er Erbium 167	69 Tm Thulium 169	70 Yb Ytterbium 173	71 Lu Lutetium 175

▲▲ **Actinide elements**

90 Th Thorium 232	91 Pa Protactinium (231)	92 U Uranium 238	93 Np Neptunium (237)	94 Pu Plutonium (242)	95 Am Americium (243)	96 Cm Curium (247)	97 Bk Berkelium (245)	98 Cf Californium (251)	99 Es Einsteinium (254)	100 Fm Fermium (253)	101 Md Mendelevium (256)	102 No Nobelium (254)	103 Lr Lawrencium (257)

Index

INDEX